Reaction Mecha

at a Gl

Reaction Mechanisms at a Glance

A stepwise approach to problem-solving in organic chemistry

MARK G. MOLONEY
BSc (Hons), PhD (Sydney), MA (Oxon), CChem, FRSC, FRACI
E.P. Abraham Fellow and Tutor in Chemistry
at St Peter's College and
University Lecturer at the University of Oxford

Blackwell
Science

Editorial Offices:
Osney Mead, Oxford OX2 0EL
25 John Street, London WC1N 2BL
23 Ainslie Place, Edinburgh EH3 6AJ
350 Main Street, Malden
MA 02148-5018, USA
54 University Street, Carlton
Victoria 3053, Australia
10, rue Casimir Delavigne
75006 Paris, France

Other Editorial Offices:
Blackwell Wissenschafts-Verlag GmbH
Kurfürstendamm 57
10707 Berlin, Germany

Blackwell Science KK
MG Kodenmacho Building
7–10 Kodenmacho Nihombashi
Chuo-ku, Tokyo 104, Japan

First published 2000

Set by Excel Typesetters Co., Hong Kong
Printed and bound in Great Britain
by MPG Books Ltd, Bodmin,
Cornwall

A catalogue record for this title is available from the British Library

ISBN 0-632-05002-0

Library of Congress
Cataloging-in-publication Data

Moloney, Mark G.
Reaction mechanisms at a glance: a stepwise approach to problem-solving in organic chemistry / Mark G. Moloney.
p. cm.
Includes index.
ISBN 0-632-05002-0
1. Organic reaction mechanisms.
I. Title.
QD502.5.M65 1999
547′.139—dc21 99-28232
CIP

DISTRIBUTORS

Marston Book Services Ltd
PO Box 269
Abingdon, Oxon OX14 4YN
(*Orders*: Tel: 01235 465500
Fax: 01235 465555)

USA
Blackwell Science, Inc.
Commerce Place
350 Main Street
Malden, MA 02148-5018
(*Orders*: Tel: 800 759 6102
781 388 8250
Fax: 781 388 8255)

Canada
Login Brothers Book Company
324 Saulteaux Crescent
Winnipeg, Manitoba R3J 3T2
(*Orders*: Tel: 204 837 2987)

Australia
Blackwell Science Pty Ltd
54 University Street
Carlton, Victoria 3053
(*Orders*: Tel: 3 9347 0300
Fax: 3 9347 5001)

For further information on Blackwell Science, visit our website:
www.blackwell-science.com

Contents

A note from the author

This book has been the result of some 10 years of teaching organic chemistry at the University of Oxford. During this time, I have seen the way students often have to struggle to come to terms not only with an enormous body of facts which they encounter in organic chemistry, but to develop an understanding of their interrelationship, all in a very short space of time during their first year at university. Of particular difficulty is the development of an ability to apply this knowledge to the solution of problems. This book is an attempt to demonstrate that there is indeed an underlying set of rules which can be grasped with a little effort, but that, more importantly, the use of these rules in a systematic fashion substantially reduces the burden on memory!

This systematic approach has been developed and refined in numerous tutorials with the students of St Catherine's and St Peter's Colleges, Oxford, which I have always found stimulating, but I would like particularly to acknowledge Dr Josephine Peach for many interesting discussions and for her helpful advice on the teaching of organic chemistry.

I would like to dedicate this book to my wife Julie and all the members of my family.

Mark Moloney

Introduction

There are many organic chemistry texts, old and new, which cover material from the fundamental to the advanced. Most texts, of course, are factually based, but students seem to find considerable difficulty in the application of this factual knowledge to the solution of problems, and all too often attempt to rely on memory alone. However, the sheer volume of material to be committed to memory presents a considerable burden, and the temptation to give up on the subject almost at the outset can be very strong. This book attempts to demonstrate that a general problem-solving strategy is indeed applicable to many of the common reactions of organic chemistry. It develops a check-list approach to problem-solving, using mechanistic organic chemistry as its basis, which is applicable in a wide variety of situations. It aims to show that logical and stepwise reasoning, in combination with a good understanding of the fundamentals, is a powerful tool to apply to the solution of problems.

Philosophy of the book

This book is not a 'fill in the box' text, nor does it have detailed explanations, but it does show how a problem can be worked through from the beginning to the end. The principal aim is to develop deductive reasoning, which the student is then able to apply to unfamiliar situations, using as a basis a standard list of reactions and their associated mechanisms. This book is intended for first- and second-year students, but will not cover the fundamentals of the subject, which are more than adequately covered in a variety of texts already. A knowledge of electron accounting, such as the octet rule and Lewis structures, and the meaning of curly arrows and arrow pushing is therefore assumed. However, this book will aim to reinforce and develop the *use* of these concepts by application of a generalized strategy to specific problems; this will be done using short, multi-step, reaction schemes. Note that, because the emphasis is on the strategy of problem-solving, only a limited range of problems will be covered, and no attempt has been made to achieve a comprehensive list of all reactions. The aim is to demonstrate that this strategy is applicable to a wide variety of situations, and therefore an exhaustive list of problems is considered to be inappropriate; in fact, this would defeat the very purpose of this book.

A novel layout is used, in which two facing pages will have the problem and answer. On the left page is the problem and the overall strategy; on the right side of this page are broad hints corresponding to each step. These hints are not intended to give a detailed explanation of the answer, but to provide a guide to the approach to arriving at the answer. The right-hand page will have a complete worked solution. Placing a piece of A4 paper on the right-hand page will both provide a working space and hide the full answer from view. The intention is that this will remove the temptation to look at the complete solution too early, but still provide access both to the stepwise procedure for working through the problem and to the hints on the left page. Of course, maximum benefit from these problems will come only if they are worked through in their entirety before looking at the worked solution! The detailed answer will be as full as possible within the page constraints and will include, for example, proton transfers. A new innovation will be introduced regarding curly arrows; these are labelled in sequence thus (a, b, c, d), to clearly indicate to the student the starting point and the subsequent sequence of movement of electrons. It is hoped that the provision of both hints and worked solutions will cater for a variety of academic abilities.

However, it should be emphasized that no matter how well a strategy for problem-solving is developed, there is no substitute for a good knowledge of the subject. One might consider that learning organic chemistry is little different from learning a foreign language. The vocabulary of any language is very important and must be learnt, and for organic chemistry these are the standard reactions of the common functional groups. A checklist of the reactions which a student is expected to know and which form the basis of the questions in that chapter is summarized before each set of problems. No detail is included, however, since there are many excellent texts available which cover the required material. Mechanisms, which might be considered to be the grammar of organic chemistry, are covered in considerable detail in this book. Experience shows that mechanisms are best learnt by repeated practice in problem-solving.

How to use the book

1 This text has been subdivided by functional groups, since this provides an instantly recognizable starting-point, especially for the beginning student. A key skill, therefore, which must be developed early, is the recognition of functional groups and recollection of their typical or characteristic reactions. This information is briefly summarized at the beginning of every chapter, but an important starting point is to prepare your own set of more detailed notes, using your recommended texts and lecture notes as source material. There is no alternative but to commit this material to memory.
2 These characteristic reactions can be very often understood using some fundamental chemical principles; mechanisms provide a way of rationalizing the conversion of starting materials to products. In order to devise plausible mechanisms (remember that the only way of verifying any postulate is by experiment), it is necessary first to be able to identify nucleophiles and electrophiles, Lowry–Brønsted and Lewis acids and bases and leaving groups.
3 A further aid to problem-solving is to number the atoms in the starting material and the corresponding atoms in the product; this allows for effective atom 'bookkeeping'.
4 In devising plausible mechanisms, it is usually most helpful to begin at an electron-rich centre (negative charge, lone pair or carbon–carbon double or triple bond) and push electrons (i.e. begin an arrow thus: ⌒) from there.
5 Remember that a double-headed arrow (⌒) refers to movement of two electrons and a single-headed arrow (⌒) to a single electron.
6 Each problem in this book is designed to illustrate a sequential strategy of thinking to solve a question of the type: 'Provide a plausible explanation of the following interconversion; in your answer, include mechanisms for each reaction step.' A typical example is:

Cl–C1–C2–OH (atoms numbered 1, 2) → MeS–CH₂CH₂–O–C(=O)–CH₂–NH–CH₂–Ph

(i) MeSNa
(ii) $SOCl_2$, py
(iii) $PhCH_2NHCH_2CO_2H$, Et_3N

Note that numbering of each reagent, but not the product, is already given at this stage. Also the sequence of reagents drawn means that the intermediate product of step (i) is subsequently treated with the reagents of step (ii), whose product in turn is treated with the reagents of step (iii) to give the final product shown.
7 The general strategy is:

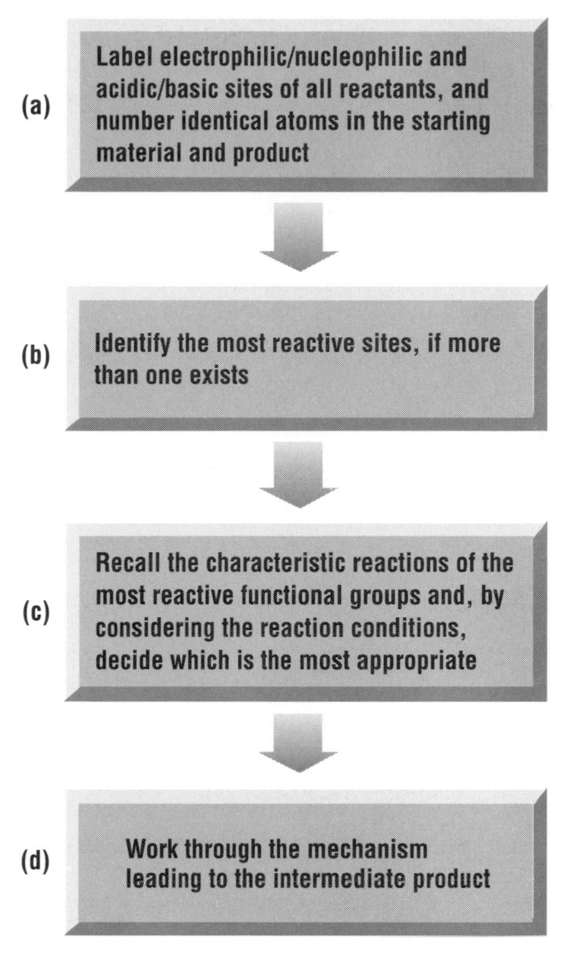

(a)
Label electrophilic/nucleophilic and acidic/basic sites of all reactants, and number identical atoms in the starting material and product
(b)
Identify the most reactive sites, if more than one exists
(c)
Recall the characteristic reactions of the most reactive functional groups and, by considering the reaction conditions, decide which is the most appropriate
(d)
Work through the mechanism leading to the intermediate product

The first step (box (a)) is an important preparatory stage, and this information is given on the right-hand answer page above the dotted line, thus:

δ- Cl δ+ 1 2 OH δ- $MeS^{\ominus}$ $Na^{\oplus}$ δ- O=S(Cl)Cl δ+   MeS 1 2 Cl

Also included in this section are any preliminary reactions necessary to generate reactive species, e.g.

$$PhCH_2NHCH_2CO_2H \quad + \quad Et_3N \rightleftharpoons PhCH_2NHCH_2CO_2^{\ominus} \quad + \quad Et_3\overset{\oplus}{N}H$$

In order to solve a problem with multiple reaction sites, it is necessary to recognize which is the most reactive (box (b)). For the example shown, alkyl halides are electrophilic at the carbon adjacent to the halogen and alcohols are nucleophilic on oxygen. Once the most reactive functional group has been identified, recollection of its characteristic reactions (box (c)) is a useful next step. In the example given, alkyl halides readily undergo nucleophilic substitution reactions; although the carbon adjacent to the oxygen is also electrophilic (by virtue of their electronegativity difference), hydroxide is a poor leaving group. Once this is established, the next step (box (d)) can be deduced. To continue with this example, nucleophilic attack of thiolate, a potent nucleophile, at the alkyl halide generates the substitution product.

Fragments which are lost in the course of these steps (e.g. leaving groups, such as H_2O or Cl^-) are indicated as '-X' under or beside the relevant reaction arrow, thus:

$$\xrightarrow[\text{-X}]{}$$

Two points are noteworthy: where possible, a series of unimolecular or bimolecular steps have been used; termolecular steps might make for a more concise answer, but are kinetically much less favourable! It is also important to remember that many reactions are in fact equilibria, and that the overall transformation of starting materials to products often crucially depends on one or more irreversible steps in a reaction sequence; where space permits, this is indicated.

8 This process can be applied for as many iterations as are necessary in any given problem; when you have come to the end of one iterative cycle, the product (which could be an intermediate one or the final one of the question) is in a box, thus:

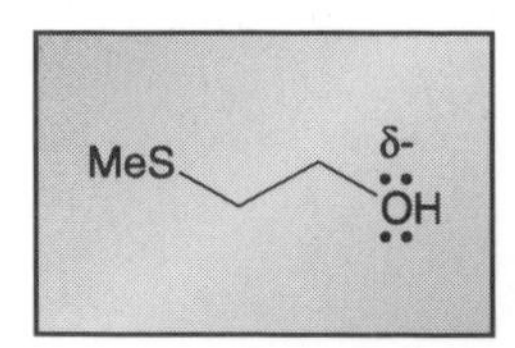

Once you have reached this stage, you will need to return to the beginning of the cycle (i.e. box (a)) and proceed through the sequence again; relevant labelling (e.g. polarization, lone pairs) of the substrate is included in the structure in the box. Note that the number of iterations required to reach the final product is different for each question; you will need to use your judgement accordingly. However, for a multi-step sequence like that in the example, you could expect to need *at least* the same number of iterations as there are steps, i.e. in this case, three. Each complete iteration is briefly explained in separate paragraphs beginning with an asterisk (*).

9 However, often a penultimate product is obtained which does not look like the desired product, but is in fact very close to it; this can be very misleading, and needs to be watched for with care. Tautomerization is a good example. Under these circumstances, the following step will be necessary:

Recognize that this is not the final product, but is closely related to it

Note that only general acidic or basic work-up conditions are indicated, and this implies that the final product can be obtained by protonation or deprotonation, respectively. If base or acid reagents are specified for any reaction sequence, a reaction (other than simple protonation or deprotonation) is implied.

10 All going well, you should now be in a position to:

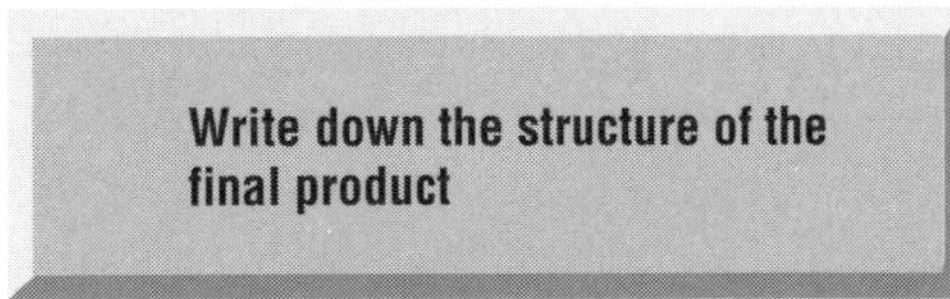

Write down the structure of the final product

11 Hints are provided adjacent to this general strategy; however, you may not need to use them and, if not, so much the better! **Bold** terms should be familiar, but if not they can be checked in any suitable textbook.

12 The detailed answer is provided on the right-hand page, allowing you to check your answer. Avoid the temptation to look at this until you have entirely finished the question!

13 There are seven questions per chapter, designed to cover as many of the full range of characteristic reactions for each functional group as possible. At the end of each chapter, there are supplementary questions, which are designed to reinforce the lessons of each detailed question.

14 Remember that this strategy is an aid to solving problems, and is not always universally applicable; all problems are different, and slavish following of this approach is no guarantee of certain success. This strategy is no substitute for thinking!

15 There is one particularly important limitation to this strategy; in this book, it is designed specifically for mechanisms involving polar intermediates (hence the emphasis on electrophilic and nucleophilic processes). The strategy is not, however, *directly* applicable to radical reactions or pericyclic reactions, and reactions of this type have therefore been largely, although not exclusively, omitted. In any case, these types of reaction are considered to be too advanced for introductory organic chemistry.

16 Obviously, this approach is designed to develop your understanding of the subject and, in the short term, to be of use in those all-important examinations. The strategy is meant to make problem-solving easier, and even fun!

List of abbreviations

Ac	Acetyl ($CH_3C(O)$–)
cat.	Catalytic
Δ	Heat
DMF	*N*,*N*-Dimethylformamide (Me_2NCHO)
MCPBA	*meta*-Chloroperbenzoic acid
PPA	Polyphosphoric acid
p-TsCl	*p*-Toluenesulfonyl chloride (*p*-$MeC_6H_4SO_2Cl$)
p-TsOH	*p*-Toluenesulfonic acid (*p*-$MeC_6H_4SO_3H$)
py	Pyridine (C_5H_5N)
THF	Tetrahydrofuran

1 Nucleophilic substitution and elimination

Nucleophilic substitution: S_N1 and S_N2 reactions

S_N1

sp2
planar

equal amounts (racemate)

S_N2

- S_N1 is stepwise and unimolecular, proceeding through an intermediate carbocation; S_N2 is bimolecular with simultaneous bond-making and bond-breaking steps, but does not proceed through an intermediate.
- The nature of the nucleophile and leaving group, solvent polarity and substrate structure can all alter the mechanistic course of substitution. The stability of carbocation intermediates in S_N1 is very important, and rearrangements can occur.
- There are important stereochemical consequences of the S_N1 and S_N2 mechanisms (racemization vs. inversion).
- Steric effects are particularly important in the S_N2 reaction (neopentyl halides).
- Neighbouring group participation in S_N1 reactions can be important.
- Special cases:
  (a) Allylic nucleophilic displacement—S_N1' and S_N2'.
  (b) Aryl (PhX) and vinylic ($R_2C{=}CRX$) halides—these are generally unreactive towards nucleophilic displacement.

Elimination: E_1 and E_2 eliminations

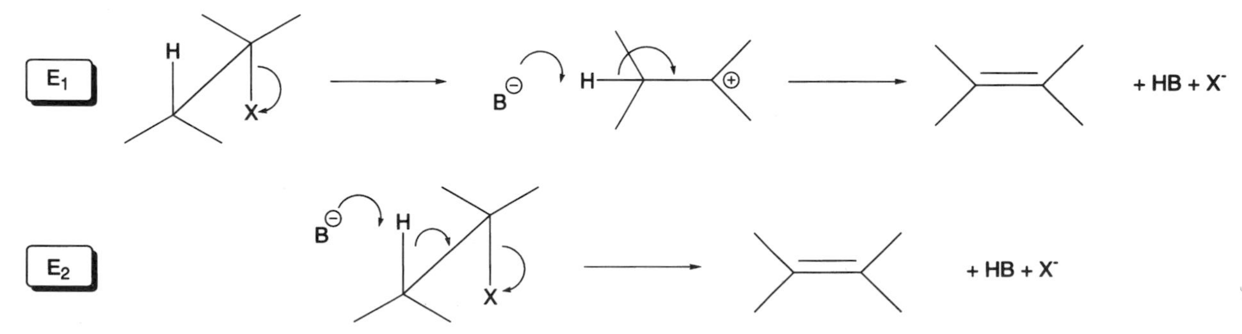

- E_1 is stepwise and unimolecular, proceeding through an intermediate carbocation; E_2 is bimolecular with simultaneous bond-making and bond-breaking steps, but does not proceed through an intermediate.
- The Saytzev and Hofmann rules can be used to predict the orientation of elimination, and the stereochemistry is preferentially antiperiplanar rather than synperiplanar due to favourable orbital overlap.
- Elimination and substitution are often competing reactions.

1.1

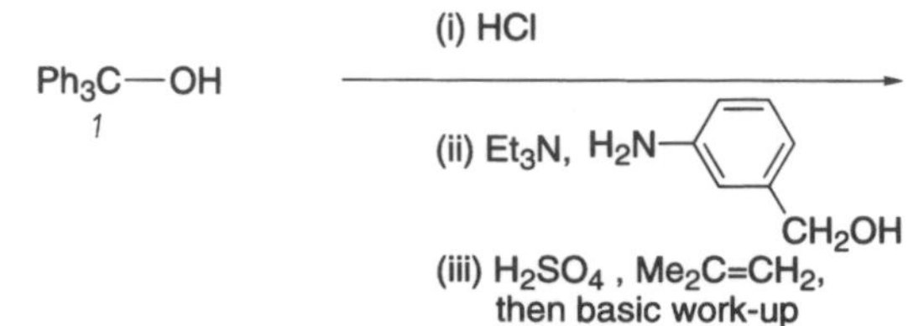

Label electrophilic/nucleophilic and acidic/basic sites of all reactants, and number identical atoms in the starting material and product

Alcohols are nucleophiles and bases, since the oxygen possesses lone pairs. HCl is a strong acid and fully ionized ($pK_a = -7$).

Identify the most reactive sites, if more than one exists

Aromatic rings can be protonated, but the alcohol is the most basic and nucleophilic site of Ph_3COH.

Recall the characteristic reactions of the most reactive functional groups and, by considering the reaction conditions, decide which is the most appropriate

Alcohols are easily protonated by strong acids, which converts the hydroxyl into a good **leaving group**, an oxonium ion, which is able to depart as water.

Work through the mechanism leading to the intermediate product

The leaving group departs to give a **resonance** stabilized triphenylmethyl cation which is then intercepted by chloride.

Repeat the above four steps...

* Triphenylmethyl chloride readily undergoes S_N1 reactions; departure of the good **leaving group** (chloride) regenerates the carbocation, which is intercepted by the most nucleophilic functional group of the aniline reagent, that is, the amine group. A series of proton transfers then gives the product.
* Under stongly acidic conditions (H_2SO_4), isobutene is protonated (**Markovnikov** addition) to give a *t*-butyl cation; this is intercepted to give the ether product in its protonated form.

Recognize that this is not the final product, but is closely related to it

Deprotonation of this **oxonium** cation gives the ether product.

Write down the structure of the final product

Summary: There are several examples of nucleophilic substitution (S_N1) reactions in this question.

$$Nu + R{-}X \longrightarrow R{-}Nu + X$$

Now try questions 1.8 and 1.9 at the end of this Chapter

1.2

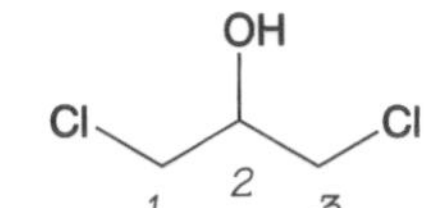

(i) $Ca(OH)_2$, H_2O

(ii) 2 MeSNa
(iii) NaH, THF then MeI, THF

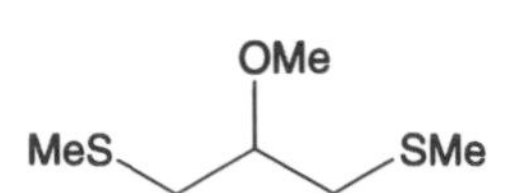

Label electrophilic/nucleophilic and acidic/basic sites of all reactants, and number identical atoms in the starting material and product

Alkyl chlorides are good electrophiles (chloride is a good leaving group and its electronegativity polarizes the C–Cl bond). Alcohols are nucleophiles and weak acids. Calcium hydroxide is a weak base.

Identify the most reactive sites, if more than one exists

Under the basic conditions of the reaction, the alcohol function is deprotonated to give an alkoxide anion, which is very nucleophilic.

Recall the characteristic reactions of the most reactive functional groups and, by considering the reaction conditions, decide which is the most appropriate

Alkyl halides readily undergo nucleophilic substitution reactions with alkoxides to give ethers (the **Williamson ether synthesis**).

Work through the mechanism leading to the intermediate product

The alkoxide undergoes an intramolecular **nucleophilic substitution** reaction to give an epoxide.

Repeat the above four steps...

* Methanethiolate is a good nucleophile, and both attacks the C–Cl bond and the epoxide function (which has two electrophilic sites) at the less hindered end, to give the alkoxide product; this is protonated on work-up.
* Sodium hydride is a good base, and deprotonates the alcohol; *alkylation* with MeI, via a nucleophilic substitution mechanism, gives the final ether product.

Recognize that this is not the final product, but is closely related to it

Not needed here.

Write down the structure of the final product

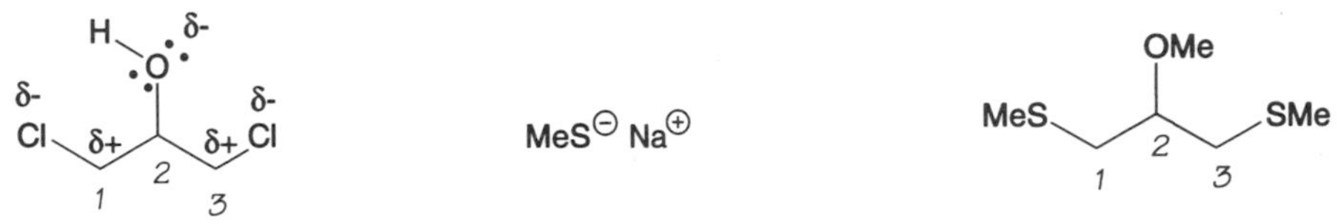

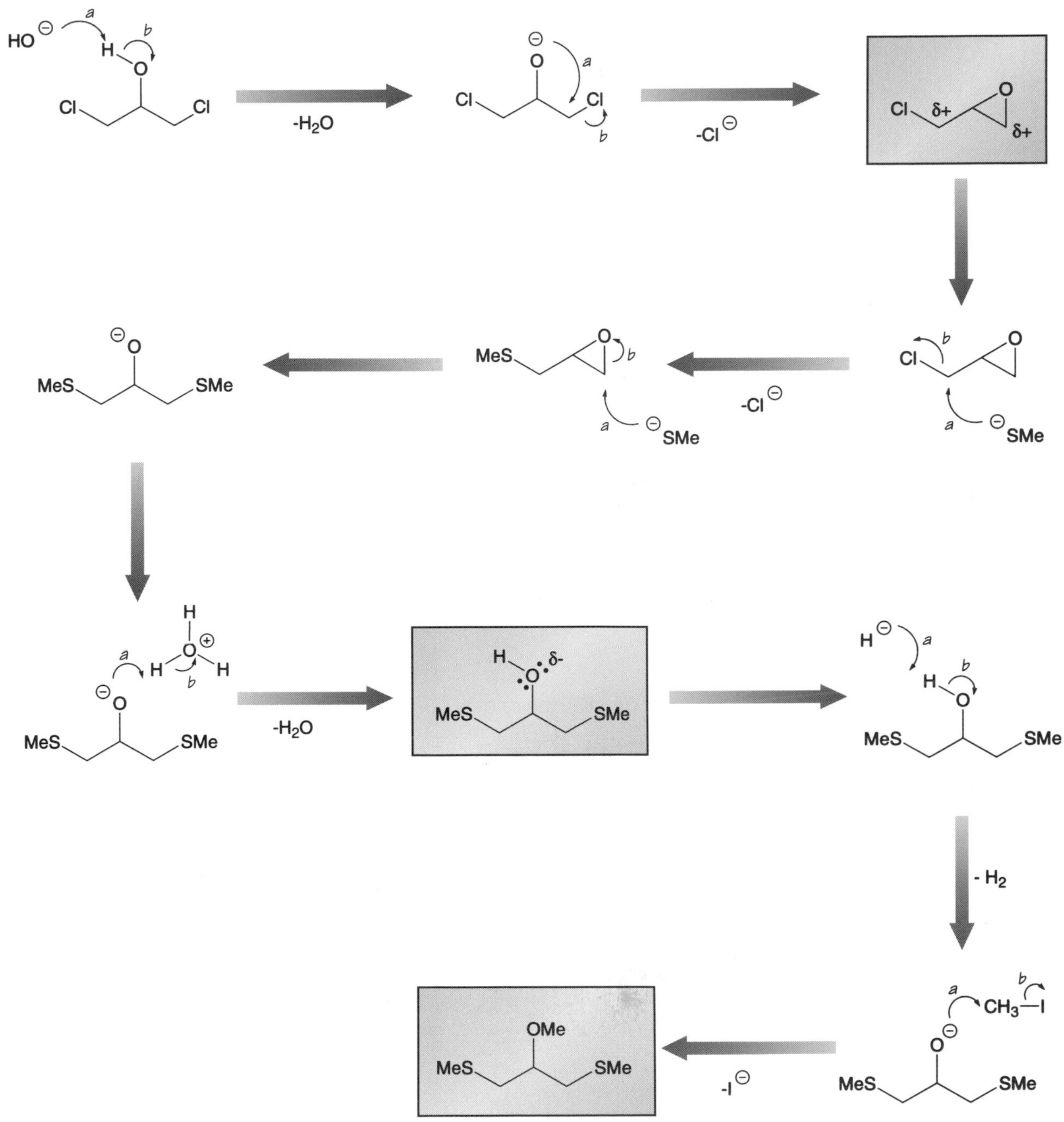

$MeS^{\ominus}\ Na^{\oplus}$

$-H_2O$

$-Cl^{\ominus}$

$-Cl^{\ominus}$

$-H_2O$

$-H_2$

$-I^{\ominus}$

Summary: This question involves several examples of nucleophilic substitution (S_N2) reactions.

$$Nu^{\ominus} + R{-}X \longrightarrow R{-}Nu + X^{\ominus}$$

Now try questions 1.10 and 1.11 at the end of this Chapter

1.3

Cl–CH₂(1)–CH₂(2)–OH → MeS–CH₂CH₂–O–C(=O)–CH₂–NH–CH₂–Ph

(i) MeSNa

(ii) $SOCl_2$, py

(iii) $PhCH_2NHCH_2CO_2H$, Et_3N

Label electrophilic/nucleophilic and acidic/basic sites of all reactants, and number identical atoms in the starting material and product

Alkyl chlorides are good electrophiles (chloride is a good leaving group) and alcohols are good nucleophiles (the oxygen has lone pairs).
Methanethiolate is an excellent nucleophile (the sulfur is very polarizable and carries a negative charge).

Identify the most reactive sites, if more than one exists

Since the reaction is with the nucleophilic reagent, MeS^-, the most reactive site is the alkyl chloride. An alcohol is not reactive with a nucleophilic reagent.

Recall the characteristic reactions of the most reactive functional groups and, by considering the reaction conditions, decide which is the most appropriate

Alkyl chlorides readily undergo **nucleophilic substitution** reactions, since they possess a leaving group, and are electrophilic by virtue of the electronegative halogen substituent.

Work through the mechanism leading to the intermediate product

Since the reaction here is between 1° alkyl chloride and a highly nucleophilic thiolate anion, an S_N2 mechanism is most likely.

Repeat the above four steps...

* Thionyl chloride is highly electrophilic, and converts the alcohol to the corresponding alkyl chloride **via** an **addition–elimination** process (with **neighbouring group** or **anchimeric** assistance of the SMe group).
* The nucleophilic hydroxyl oxygen of the carboxylic acid undergoes a **nucleophilic substitution** reaction with the alkyl chloride (with **anchimeric** assistance of the SMe group) to give a protonated ester.

Recognize that this is not the final product, but is closely related to it

This penultimate product is then deprotonated by Et_3N to give the ester product.

Write down the structure of the final product

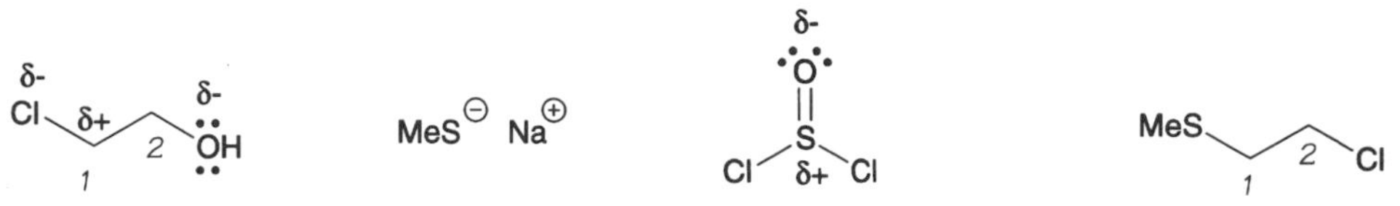

$PhCH_2NHCH_2CO_2H$ + Et_3N ⇌ $PhCH_2NHCH_2CO_2^{\ominus}$ + $Et_3\overset{\oplus}{N}H$

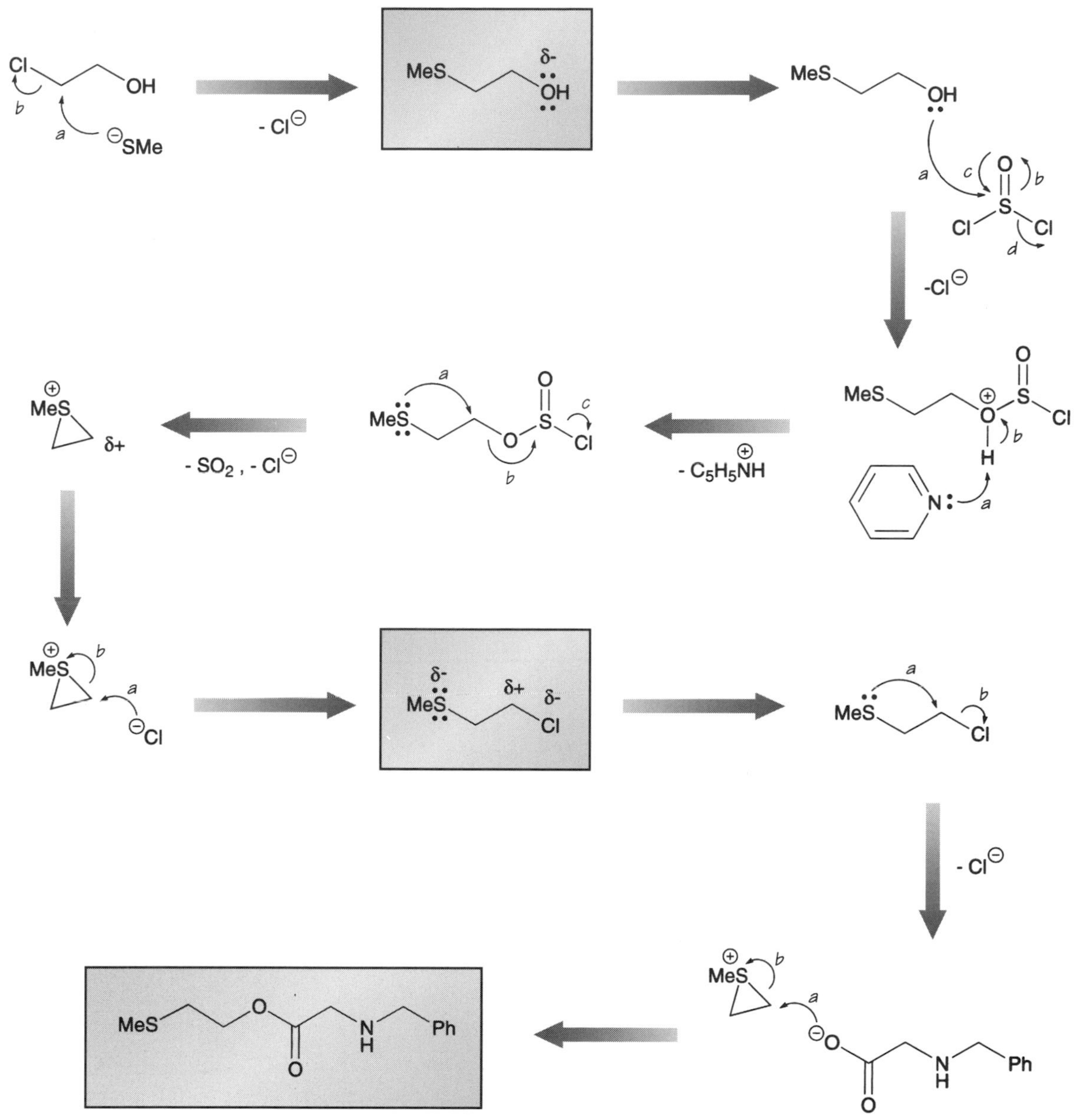

Summary: *This is an example of anchimeric assistance (also called neighbouring group participation) in nucleophilic substitution reactions.*

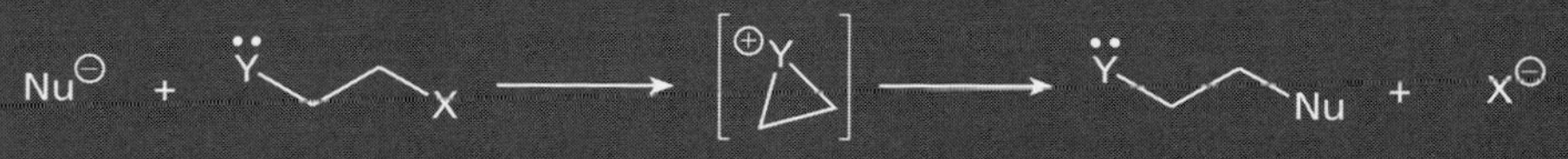

$Nu^{\ominus}$ + Y–CH₂CH₂–X ⟶ [$\oplus$Y three-membered ring] ⟶ Y–CH₂CH₂–Nu + $X^{\ominus}$

Now try questions 1.12 and 1.13 at the end of this Chapter

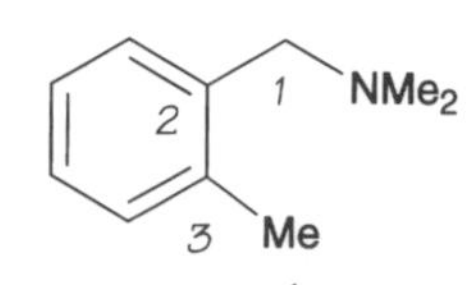

(i) EtBr, EtOH, Δ
(ii) NaOAc, AcOH, Δ
(iii) NaOH, Δ, then acidic work-up
(iv) 2 BuLi, then EtBr, then acidic work-up

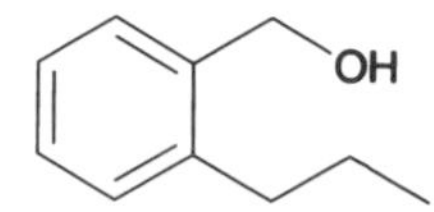

Label electrophilic/nucleophilic and acidic/basic sites of all reactants, and number identical atoms in the starting material and product

Amines are good nucleophiles (the nitrogen has a lone pair).
Alkyl bromides are electrophiles, since bromine is electronegative and bromide is a good leaving group.

Identify the most reactive sites, if more than one exists

Although an aromatic ring is a nucleophile, the most nucleophilic site of this molecule is the dimethylamino group.

Recall the characteristic reactions of the most reactive functional groups and, by considering the reaction conditions, decide which is the most appropriate

Alkyl halides readily undergo **nucleophilic substitution** reactions, and in this case the nucleophile is the amine function.

Work through the mechanism leading to the intermediate product

Since bromoethane is a 1° alkyl halide and amines are good nucleophiles, the reaction will proceed by an S_N2 process, to give a quaternary ammonium salt.

Repeat the above four steps...

* Quaternary ammonium groups are good leaving groups, and are easily displaced by the nucleophile acetate, giving in this case an acetate ester.
* Esters are easily hydrolysed under alkaline conditions (**addition–elimination** mechanism) to give the alcohol product.
* Reaction with butyl lithium (a strong base) firstly deprotonates the more acidic alcohol, and only then deprotonates the benzylic methyl group to give a resonance stabilized carbanion; this nucleophilic carbanion is quenched with bromoethane (S_N2 reaction).

Recognize that this is not the final product, but is closely related to it

The penultimate alkoxide product is protonated on acidic work-up.

Write down the structure of the final product

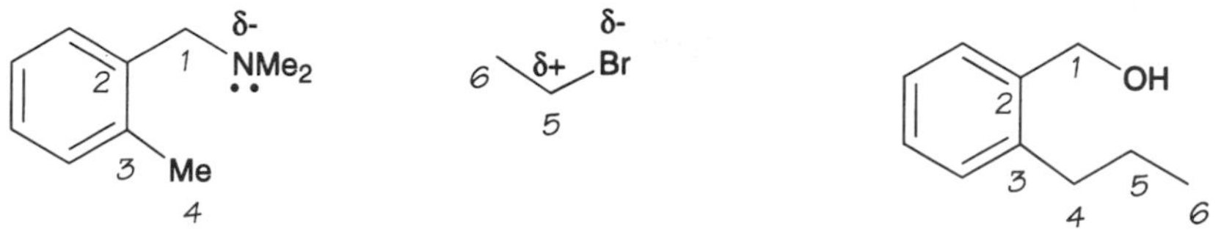

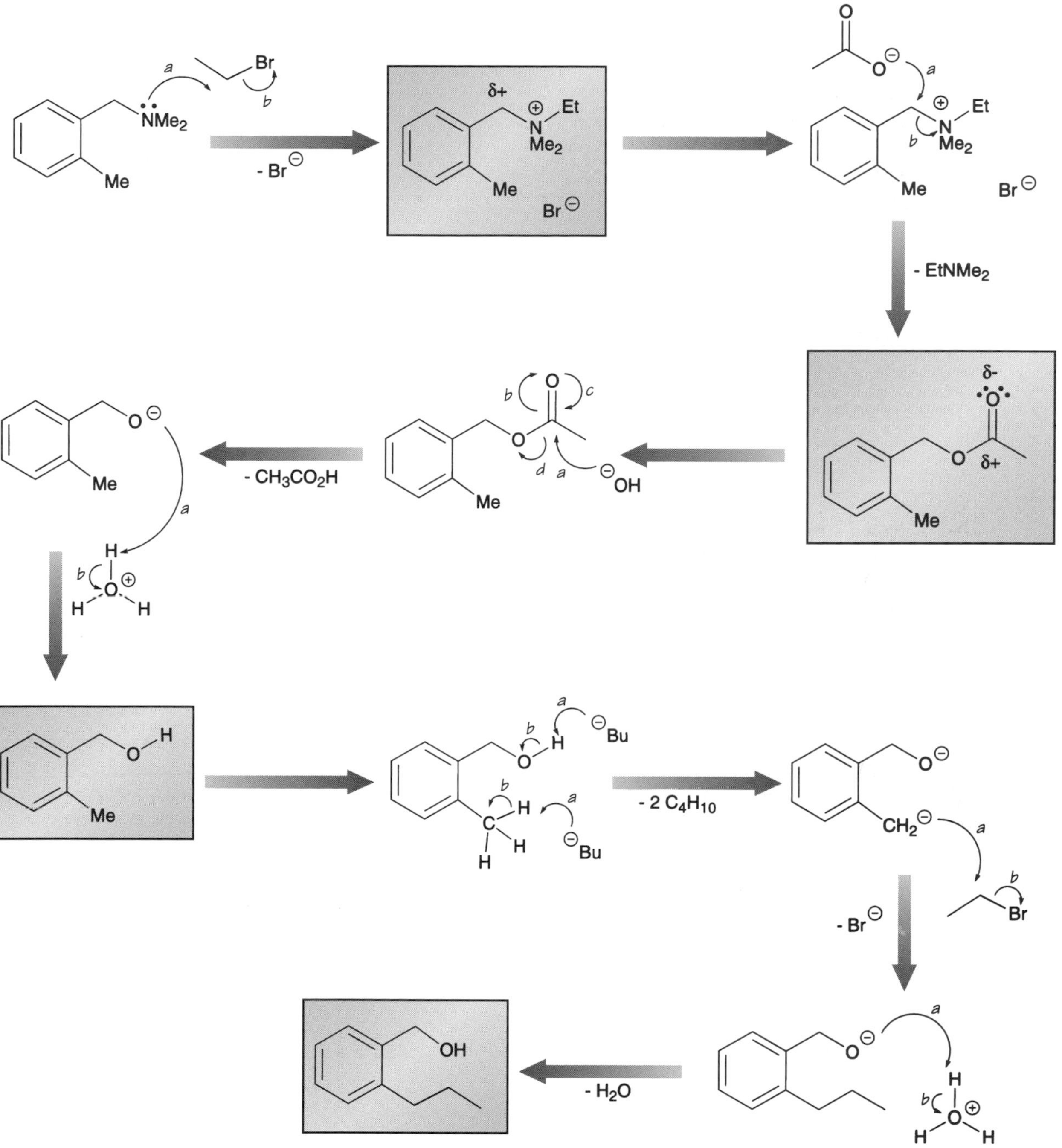

- CH_3CO_2H

- $EtNMe_2$

- $Br^{\ominus}$

- 2 C_4H_{10}

- H_2O

Summary: This question includes an example of nucleophilic attack at the benzylic position.

$$PhCH_2X + Nu^{\ominus} \longrightarrow PhCH_2Nu + X^{\ominus}$$

Now try questions 1.14 and 1.15 at the end of this Chapter

1.5

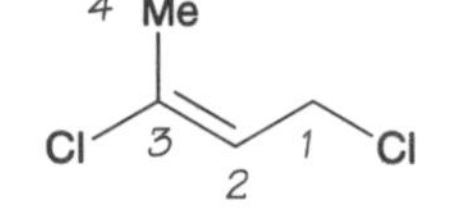

(i) Na_2CO_3, H_2O
(ii) 2 $NaNH_2$, NH_3, then acidic work-up
(iii) $SOCl_2$, py
(iv) $PhC(O)N^{-}Ph\ Na^{+}$

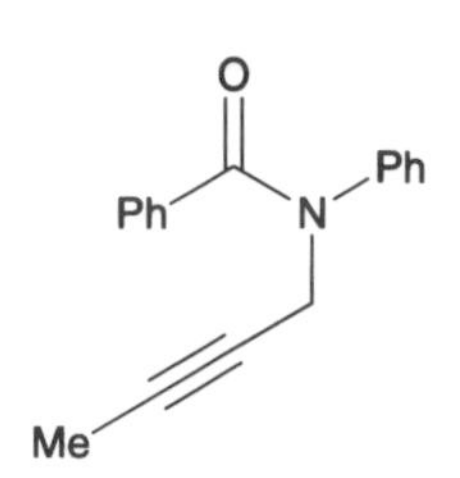

Label electrophilic/nucleophilic and acidic/basic sites of all reactants, and number identical atoms in the starting material and product

Alkyl chlorides are electrophiles, since chlorine is electronegative and chloride is a good leaving group. Sodium carbonate is a weak base; in aqueous solutions, hydroxide is generated which is a good nucleophile.

Identify the most reactive sites, if more than one exists

The **allylic** chloride is the most electrophilic site (**vinylic** chlorides have a much stronger C–Cl bond).

Recall the characteristic reactions of the most reactive functional groups and, by considering the reaction conditions, decide which is the most appropriate

Allylic chlorides are very susceptible to **nucleophilic substitution** reactions, which can be either by an S_N1 or S_N2 mechanism, depending on the solvent.

Work through the mechanism leading to the intermediate product

Allylic chlorides easily undergo S_N1 reactions in polar solvents, proceeding *via* the highly **resonance** stabilized allyl cation; this is intercepted by water, which after deprotonation gives the allyl alcohol product.

Repeat the above four steps...

* Sodamide is a strong base; **the first equivalent** deprotonates the alcohol function; the second then induces *elimination* of the vinylic chloride to give the alkyne product.
* Thionyl chloride is highly electrophilic, and converts the alcohol to the corresponding propargyl chloride *via* an **addition–elimination** process.
* **Nucleophilic substitution** by the sodium salt of PhC(O)NHPh (N more nucleophilic than O) gives the product directly.

Recognize that this is not the final product, but is closely related to it

Not needed here.

Write down the structure of the final product

$$H_2O + Na_2CO_3 \rightleftharpoons HO^{\ominus} + NaHCO_3$$

Summary: This question gives more examples of nucleophilic substitution and elimination reactions.

Now try questions 1.16 and 1.17 at the end of this Chapter

1.6

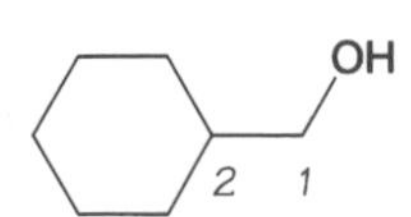

(i) HI
(ii) Me_3N
(iii) Ag_2O, H_2O then 160°C
(iv) H_2, Pd-C

[cyclohexane]–Me

Label electrophilic/nucleophilic and acidic/basic sites of all reactants, and number identical atoms in the starting material and product

Alcohols are nucleophiles and bases, since the oxygen possesses lone pairs. HI is a strong acid, and fully ionized (pK_a = −10).

Identify the most reactive sites, if more than one exists

The alcohol is the only basic and nucleophilic site in this molecule.

Recall the characteristic reactions of the most reactive functional groups and, by considering the reaction conditions, decide which is the most appropriate

Alcohols are readily protonated by acids, thereby converting the hydroxyl group into an **oxonium** ion, which is able to depart as water; they therefore readily undergo **nucleophilic substitution** reactions under acidic conditions.

Work through the mechanism leading to the intermediate product

The **oxonium** ion is a good leaving group, and is easily displaced by the good nucleophile, iodide, in an S_N2 process (1° substrate, good nucleophile).

Repeat the above four steps...

* Trimethylamine is a good base and nucleophile. Alkyl iodides are also reactive to S_N2 reactions, and the iodide is easily displaced to give a quaternary ammonium iodide.
* Treatment with silver oxide converts the iodide to the hydroxide salt (driven by the precipitation of AgI); heating of this salt causes an **elimination** reaction (**Hofmann**) to give the corresponding alkene.
* Hydrogenation of the alkene using Pd supported on charcoal as catalyst gives the alkane (syn-addition of H_2).

Recognize that this is not the final product, but is closely related to it

Not needed here.

Write down the structure of the final product

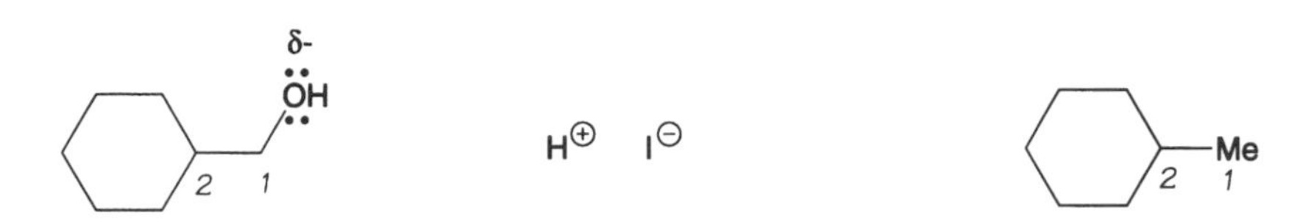
δ-
OH
2 1
H⊕ I⊖
Me
2 1

- $I^{\ominus}$

- H_2O

Me_3N

- $I^{\ominus}$

- $AgI_{(s)}$

NMe_3 $HO^{\ominus}$

- H_2O, $-NMe_3$

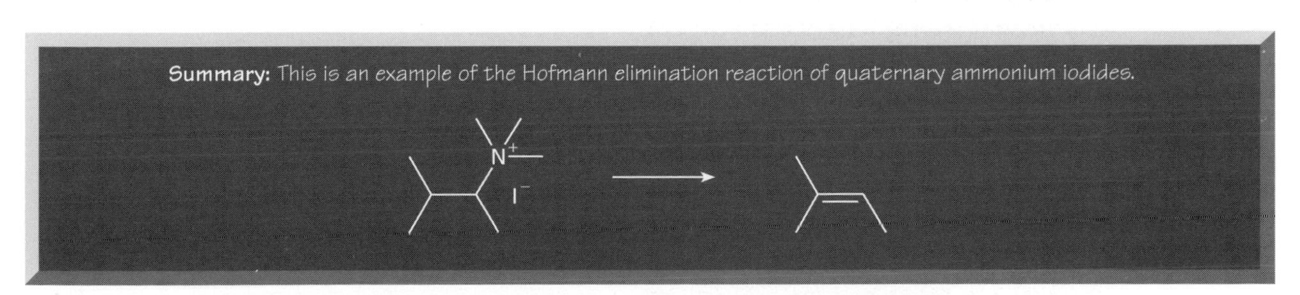
Summary: This is an example of the Hofmann elimination reaction of quaternary ammonium iodides.
N+
I−

Now try questions 1.18 and 1.19 at the end of this Chapter

(i) MeI
(ii) 0.01N NaOH/H_2O, then acidic work-up
(iii) TFA, then aqueous work-up

Label electrophilic/nucleophilic and acidic/basic sites of all reactants, and number identical atoms in the starting material and product

Sulfur, oxygen and nitrogen are all nucleophilic, since all possess lone pairs. Methyl iodide is a good electrophile, since iodine is electronegative and iodide is a good leaving group.

Identify the most reactive sites, if more than one exists

Sulfur is the most nucleophilic heteroatom since it is the least electronegative of O, N and S.

Recall the characteristic reactions of the most reactive functional groups and, by considering the reaction conditions, decide which is the most appropriate

Alkyl halides readily undergo **nucleophilic substitution** reactions, the nucleophile in this case being the sulfur atom.

Work through the mechanism leading to the intermediate product

The nucleophilic sulfur undergoes a *nucleophilic substitution* reaction with methyl iodide (S_N2) to generate a *sulfonium* cation.

Repeat the above four steps...

* The α-protons of the sulfonium cation are very acidic (highly stabilized conjugate base), and only weak base is required for deprotonation; this induces an **elimination** reaction.

* Esters can be protonated on the carbonyl oxygen by strong acids (e.g. CF_3CO_2H); the ester alkyl group departs in an E_1 process, to give the amino acid product and a *t*-butyl cation. The cation is intercepted by water, to give an **oxonium** cation.

Recognize that this is not the final product, but is closely related to it

Deprotonation of the **oxonium** cation on work-up gives the product, *t*-butyl alcohol.

Write down the structure of the final product

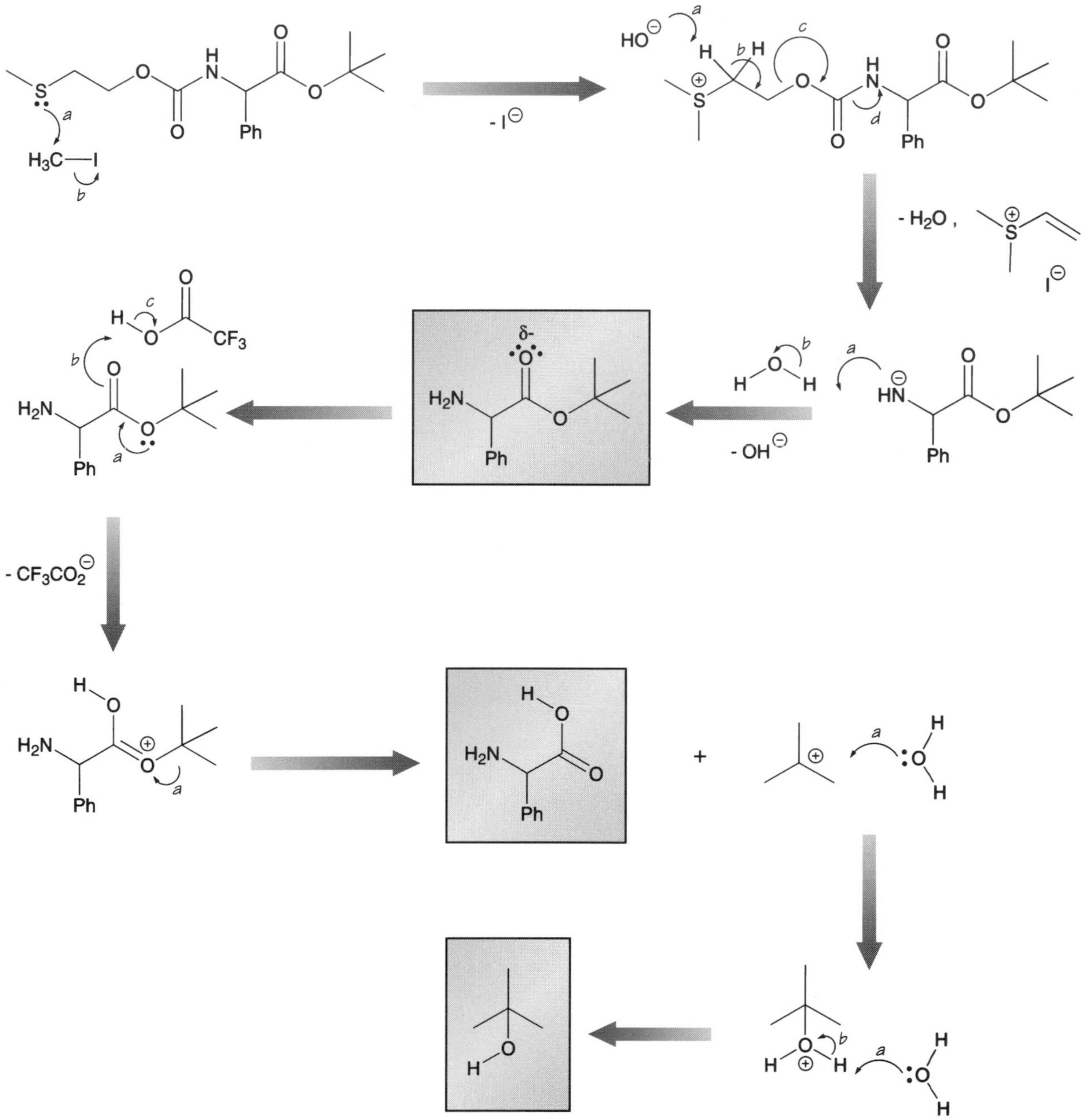

Summary: This is an example of a base catalysed β-elimination and acid-catalysed ester hydrolysis.

Now try questions 1.20 and 1.21 at the end of this Chapter

Supplementary questions

1.8

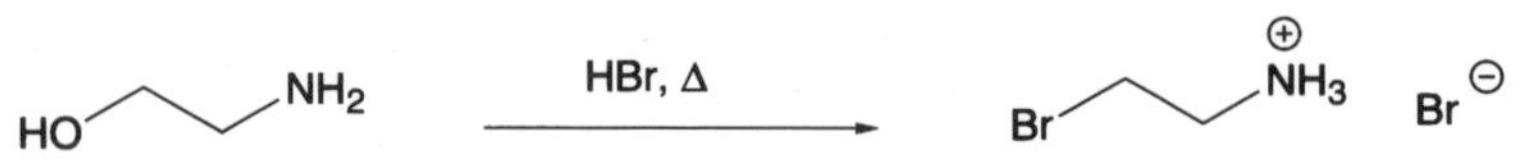

HO–CH2CH2–NH2 → (HBr, Δ) → Br–CH2CH2–NH_3^+ Br^-

1.9

(tetrahydrofuran) → (HCl, Δ) → Cl–(CH2)4–OH

1.10

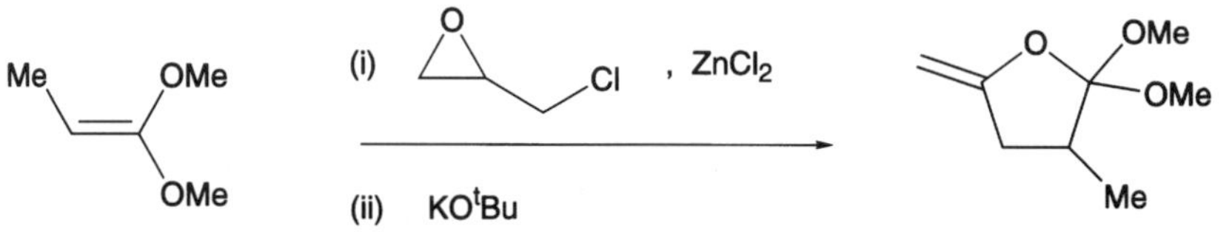

Me–CH=C(OMe)2 → (i) epichlorohydrin, $ZnCl_2$; (ii) KO^tBu → 2,2-dimethoxy-3-methyl-5-methylenetetrahydrofuran

1.11

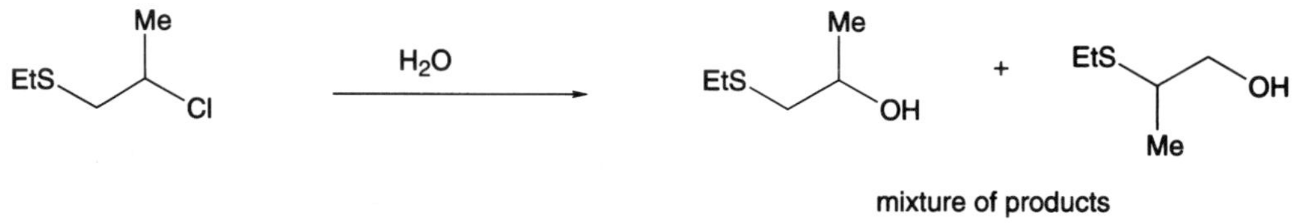

EtS–CH2–CH(Me)–Cl → (H_2O) → EtS–CH2–CH(Me)–OH + EtS–CH(Me)–CH2–OH

mixture of products

1.12

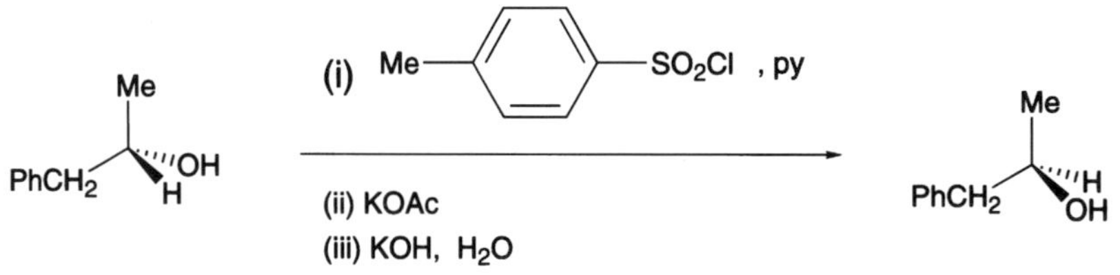

$PhCH_2$CH(Me)OH → (i) Me–C6H4–SO_2Cl, py; (ii) KOAc; (iii) KOH, H_2O → $PhCH_2$CH(Me)OH (inverted configuration)

1.13

Cl–CH2–C(Me)=CH–CH2–OAc → (i) $PhCO_2H$, Et_3N; (ii) NaOH, H_2O; (iii) PBr_3; (iv) NaI, Me_2CO → $PhCO_2$–CH2–C(Me)=CH–CH2–I

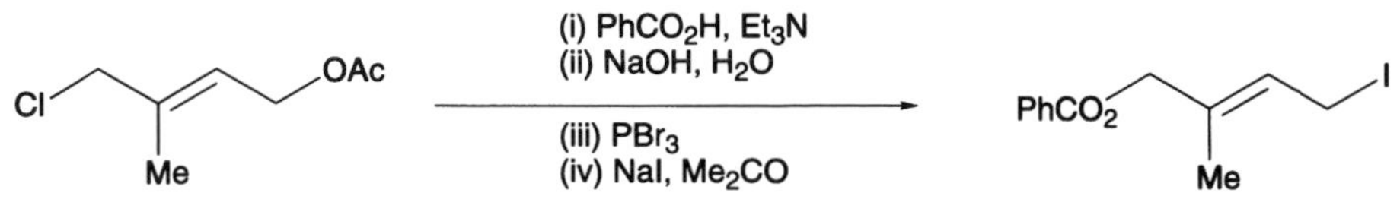

1.14

NH

(i) KH

(ii) 0.5 [epichlorohydrin] Cl

then acidic work-up

OH

1.15

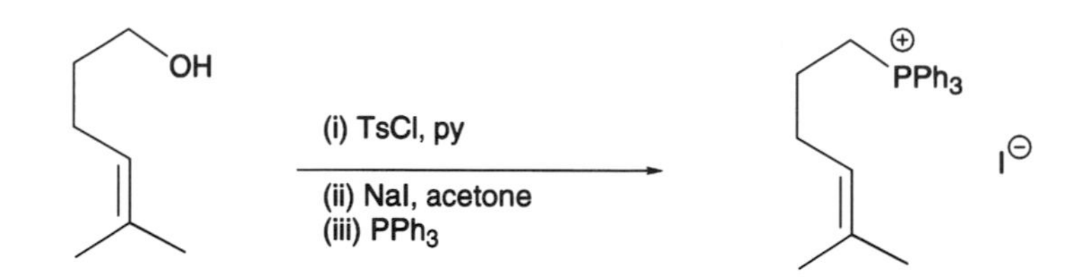

1.16

(i) $CH_2{=}C(Br)CH_2Br$
then basic work-up

(ii) $NaNH_2$, NH_3

1.17

(i) KO^tBu

(ii) [N-bromosuccinimide], $(PhCO_2)_2$

(iii) AgOAc

1.18

NH_2

(i) excess MeI

(ii) NaOEt

1.19

CH_3I

(i) [N-methylpiperidine]

(ii) AgOH

(iii) 100-200°C

1.20

MeO–CH$_2$–C≡C–H → (i) n-BuLi; (ii) Br(CH$_2$)$_4$Cl; (iii) NaCN → MeO–CH$_2$–C≡C–(CH$_2$)$_4$–CN

1.21

Epoxide–CH$_2$–CH=CH–(CH$_2$)$_3$–CO$_2$Me → (i) KBr then acidic work-up; (ii) CrO_3, H_2SO_4; (iii) Ph_3P → $Ph_3P^{\oplus}$–CH$_2$–C(=O)–CH$_2$–CH=CH–(CH$_2$)$_3$–CO$_2$Me $Br^{\ominus}$

2 Alkene and alkyne chemistry

Alkene and alkyne bonding

The molecular orbital description of bonding in alkenes (sp^2 hybridized carbon) and alkynes (sp hybridized carbon) uses π bonds.

Electrophilic addition

1 Addition of halogens:
- Mechanism and evidence for a stepwise process.
- Nature of intermediates (cyclic bromonium and iodonium species).
- Effect of substituents on reaction rate.
- Stereochemistry of addition is *anti*-.

2 Addition of hydrogen halides:
- Mechanism, and intermediacy of carbocations.
- Orientation (Markovnikov's rule), and the rationale for the effect of substitution on the orientation of addition.

3 Addition of hydrogen halides to conjugated dienes:
- 1,2- versus 1,4- addition.

Nucleophilic addition

Nucleophilic addition to C=C bonds conjugated to a C=O.

Other types of additions

1 Catalytic hydrogenation:
- Mechanism and stereochemistry—*cis*- addition of H_2.
- Nature of the catalyst.

2 Dihydroxylation:
- Osmium tetroxide or alkaline permanganate gives *syn*-1,2-diols. These products can be cleaved to dicarbonyls by treatment with periodic acid or lead tetraacetate.

3 Epoxidation with peracids.

4 Ozonolysis with ozone.

5 Diels–Alder reaction:
- Concerted reaction of dienes and dienophiles (α,β-unsaturated carbonyl compounds).

6 Hydroboration:
- Formation of a trialkylborane by *anti*-Markovnikov addition.
- Conversion of the trialkylborane to an alcohol using alkaline hydrogen peroxide.

Reaction at the allylic position

Halogenation at the allylic position using *N*-bromosuccinimide and a radical initiator (e.g. benzoyl peroxide $(PhCO_2)_2$).

2.1

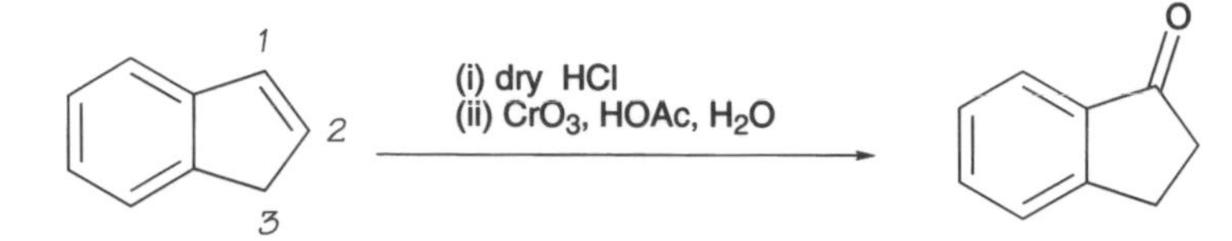

Label electrophilic/nucleophilic and acidic/basic sites of all reactants, and number identical atoms in the starting material and product

Alkenes are good nucleophiles by virtue of the π electron density.
HCl is a strong acid.

Identify the most reactive sites, if more than one exists

The cyclopentenyl double bond is the most reactive, since the double bonds of the phenyl ring are stabilized by **aromaticity**.

Recall the characteristic reactions of the most reactive functional groups and, by considering the reaction conditions, decide which is the most appropriate

Addition of HCl across the double bond, with the orientation predicted from **Markovnikov's rule**.

Work through the mechanism leading to the intermediate product

Addition of H^+ to C-2 gives a stabilized **benzylic** carbocation, which is intercepted by Cl^-.

Repeat the above four steps...

* Chloride is a good leaving group and can be displaced by the nucleophile, water, in an S_N1 process.
* The alcohol thus produced undergoes nucleophilic addition to chromium trioxide, to give a **chromate ester**, and in the process generates an excellent **leaving group** on oxygen.

Recognize that this is not the final product, but is closely related to it

Collapse of the **chromate ester** occurs to give the ketone product.

Write down the structure of the final product

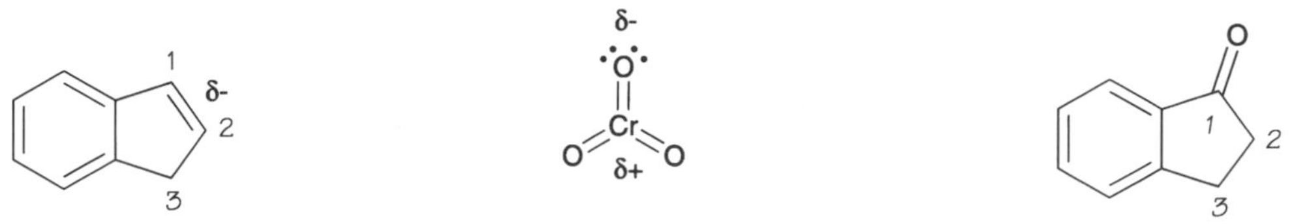

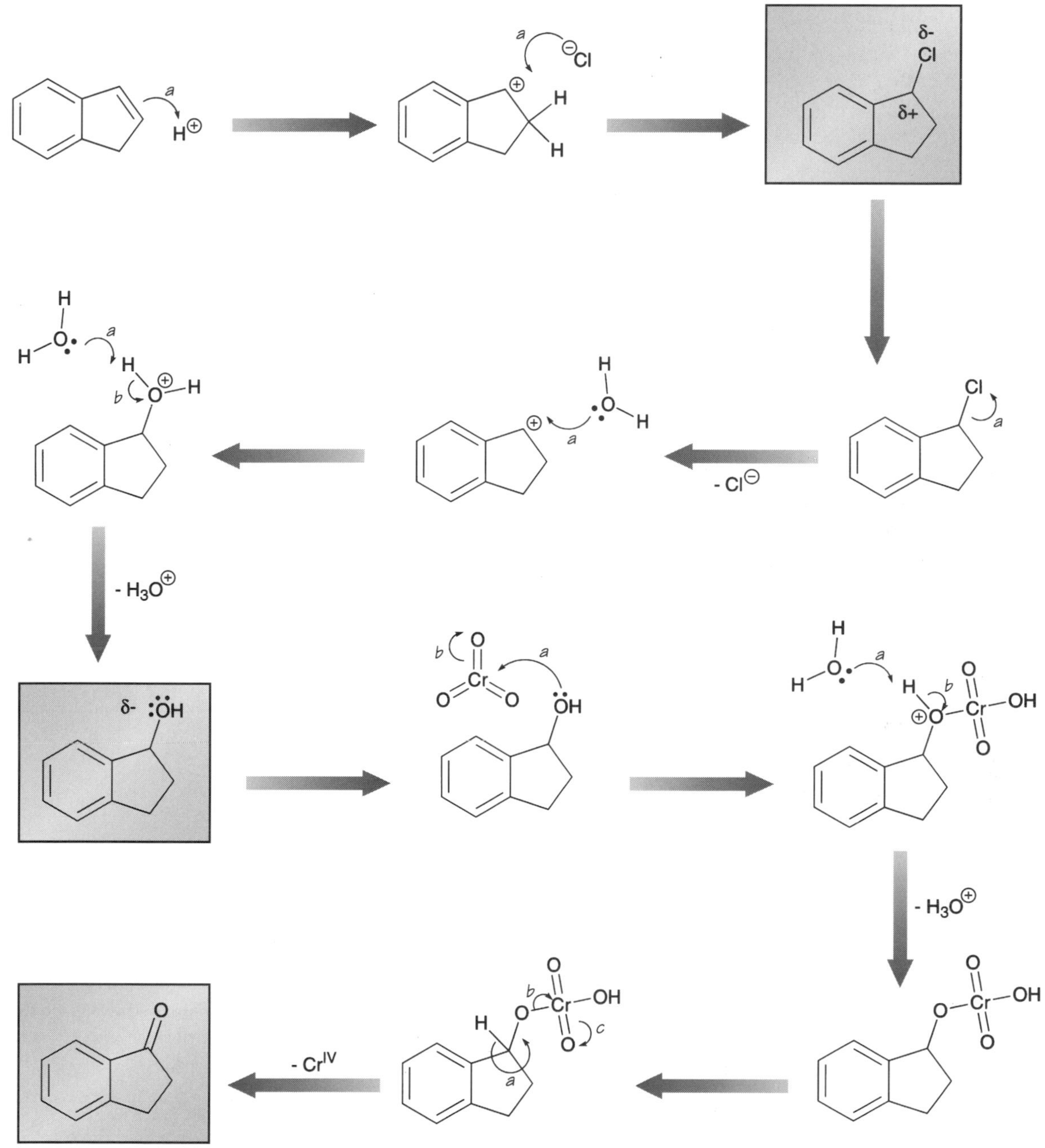

1 δ- 2 3

δ- O Cr O O δ+

O 1 2 3

a H⊕

a ⊖Cl ⊕ H H

δ- Cl δ+

Cl a

H O H ⊕ a

- Cl⊖

H H O a H b O⊕ H

- $H_3O^⊕$

δ- :ÖH

b O Cr O O a ÖH

H H O a H b ⊕O O Cr OH O

- $H_3O^⊕$

O O Cr OH O

H O b O Cr OH O c a

- Cr^{IV}

O

Summary: This is an example of Markovnikov addition to alkenes.

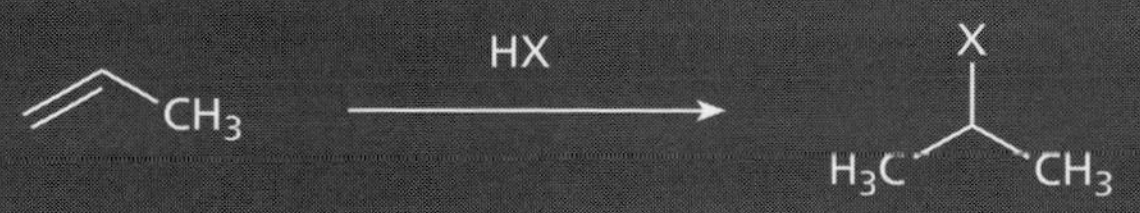

$$CH_2{=}CH{-}CH_3 \xrightarrow{HX} H_3C{-}CH(X){-}CH_3$$

Now try questions 2.8 and 2.9 at the end of this Chapter

2.2

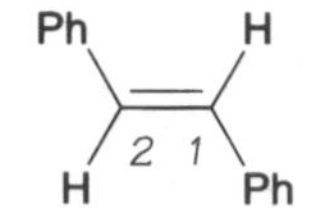

(i) Br_2, H_2O, DMSO

(ii) NaOH, EtOH
(iii) NaN_3, then acidic work-up

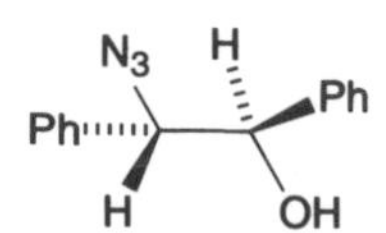

Label electrophilic/nucleophilic and acidic/basic sites of all reactants, and number identical atoms in the starting material and product

Bromine is an electrophile, with a weak Br–Br bond.
Alkenes are nucleophiles by virtue of the π electron density.

Identify the most reactive sites, if more than one exists

The double bonds of the phenyl rings are unreactive due to **aromaticity**, leaving the alkene as the most reactive nucleophile.

Recall the characteristic reactions of the most reactive functional groups and, by considering the reaction conditions, decide which is the most appropriate

Electrophilic addition of Br_2.

Work through the mechanism leading to the intermediate product

Addition of Br_2 gives an intermediate **bromonium** ion, which is intercepted by water to give a **bromohydrin**; the overall addition is *anti*-.

Repeat the above four steps...

* Treatment of the bromohydrin with base gives the corresponding alkoxide, which undergoes an intramolecular **nucleophilic substitution**, to give an epoxide, with inversion of stereochemistry at C-1.
* **Nucleophilic substitution** (S_N2) with azide occurs with inversion of stereochemistry.

Recognize that this is not the final product, but is closely related to it

Protonation of the alkoxide anion gives the product, which needs to be redrawn to show its identity with the product structure.

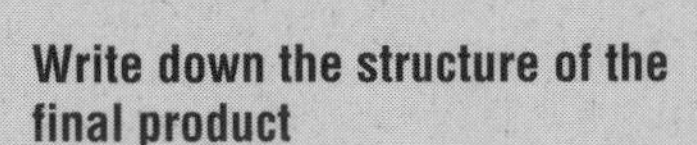

Write down the structure of the final product

$- Br^{\ominus}$

$- H_3O^{\oplus}$

$- H_2O$

$- Br^{\ominus}$

$- H_2O$

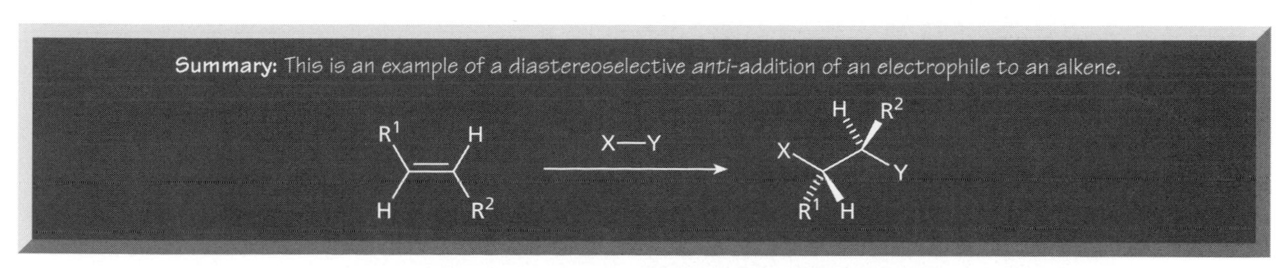

Summary: This is an example of a diastereoselective anti-addition of an electrophile to an alkene.
X—Y

Now try questions 2.10 and 2.11 at the end of this Chapter

2.3

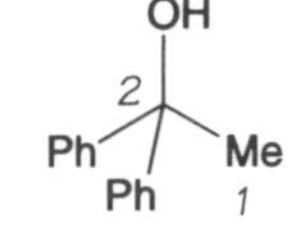

(i) dil H_2SO_4, Δ
(ii) $PhCO_3H$
(iii) CF_3CO_2H, H_2O_2
(iv) KOH, H_2O

→ PhC(=O)Ph + $H_2C{=}O$

Label electrophilic/nucleophilic and acidic/basic sites of all reactants, and number identical atoms in the starting material and product

Alcohols are nucleophiles and bases, since the oxygen possesses lone pairs. Sulfuric acid is a strong acid, and fully ionized (pK_a = −9).

Identify the most reactive sites, if more than one exists

Although an aromatic ring can be protonated, the alcohol is the most basic site.

Recall the characteristic reactions of the most reactive functional groups and, by considering the reaction conditions, decide which is the most appropriate

Alcohols readily eliminate to alkenes under acidic conditions.

Work through the mechanism leading to the intermediate product

Protonation of the alcohol gives an **oxonium** ion, which is an excellent **leaving group**, and **elimination** to the alkene occurs; the reaction is E_1, proceeding through a **benzylic** carbocation.

Repeat the above four steps...

* Perbenzoic acid is a source of electrophilic oxygen, and easily **epoxidizes** the alkene double bond.
* Trifluoroacetic acid protonates the epoxide, which then opens (via the more stabilized **benzylic** carbocation in an S_N1 process); this cation is intercepted by hydrogen peroxide.
* Hydroxide deprotonates the alcohol group.

Recognize that this is not the final product, but is closely related to it

Fragmentation of the alkoxide gives the products, via cleavage of the weak O–O bond.

Write down the structure of the final product

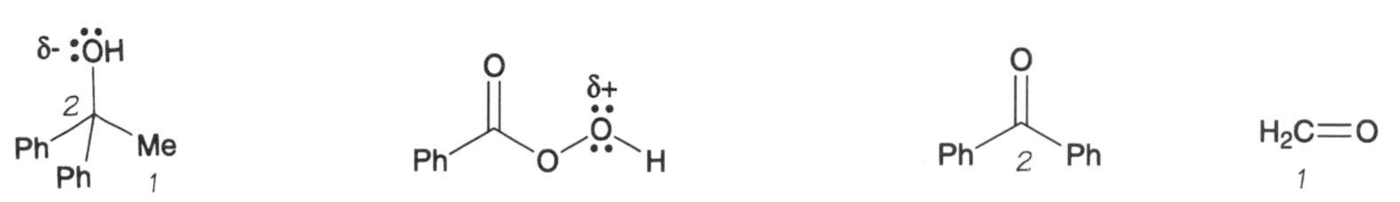

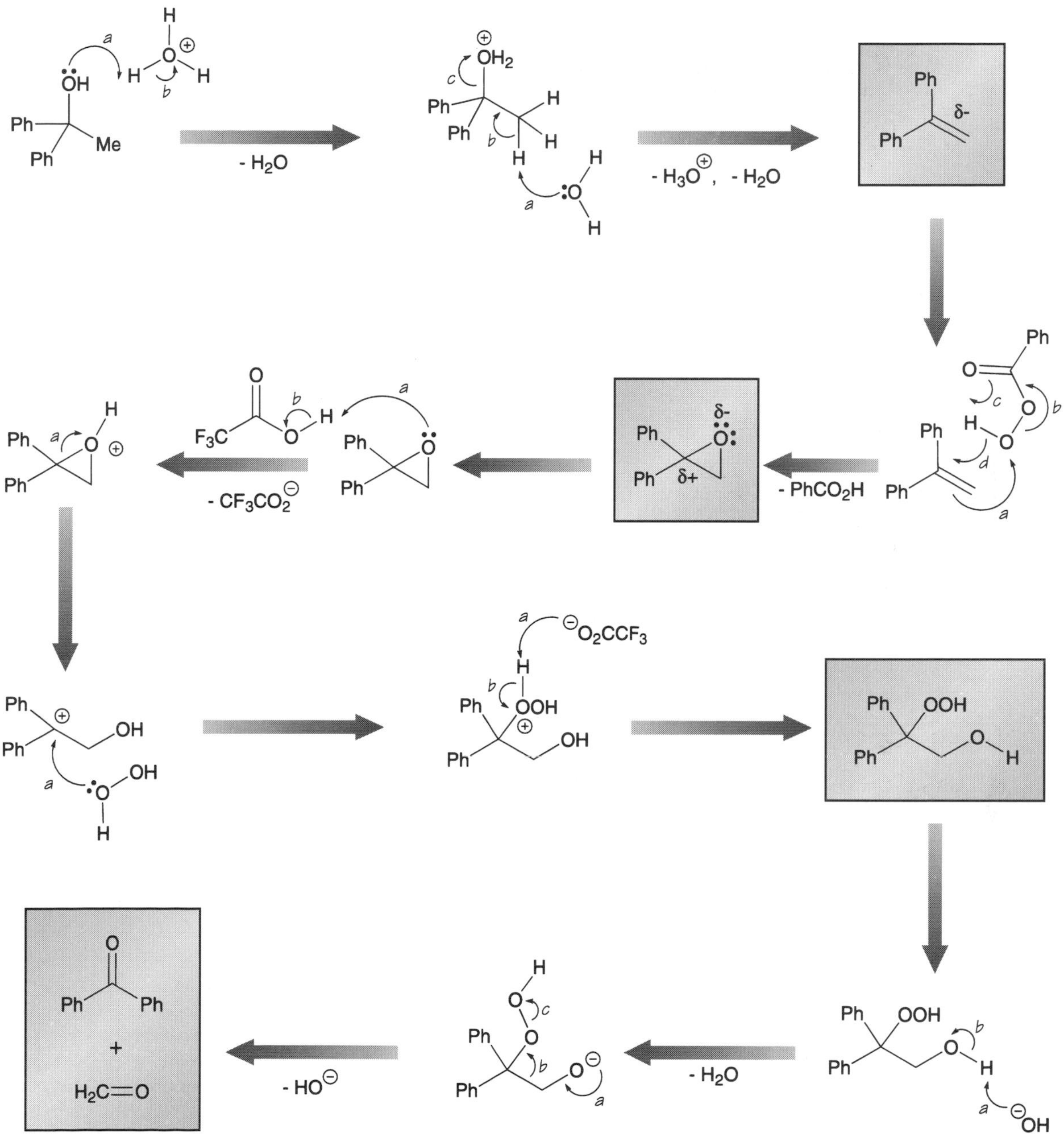

Summary: This is an example of nucleophilic substitution reactions, and elimination and addition reactions of alkenes.

Now try questions 2.12 and 2.13 at the end of this Chapter

2.4

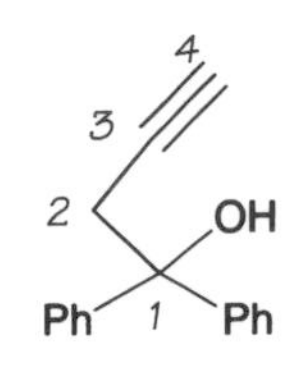

conc H_2SO_4, H_2O, HOAc

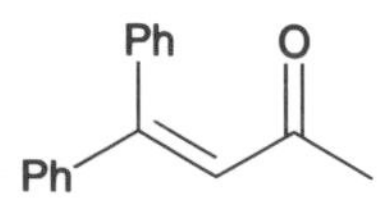

Label electrophilic/nucleophilic and acidic/basic sites of all reactants, and number identical atoms in the starting material and product

Both the phenyl and alkyne groups, and the alcohol, are nucleophilic and basic, due to the π electron density and lone pairs, respectively.
Sulfuric acid is a strong acid and fully ionized ($pK_a = -9$).

Identify the most reactive sites, if more than one exists

Although the phenyl ring and the alkyne could be protonated, the alcohol is the most basic site.

Recall the characteristic reactions of the most reactive functional groups and, by considering the reaction conditions, decide which is the most appropriate

Alcohols readily eliminate under acidic conditions to give alkenes.

Work through the mechanism leading to the intermediate product

Protonation of the alcohol to give an **oxonium** cation occurs; this is a good *leaving group*, and departure generates a stabilized **benzylic** carbocation. **Elimination** then occurs to the alkene product.

Repeat the above four steps...

Alkynes undergo **electrophilic addition** reactions. Protonation of the alkyne at the terminal position gives the more substituted **vinylic** cation, which is intercepted by water.

Recognize that this is not the final product, but is closely related to it

The resulting enol **tautomerizes** to the ketone product.

Write down the structure of the final product

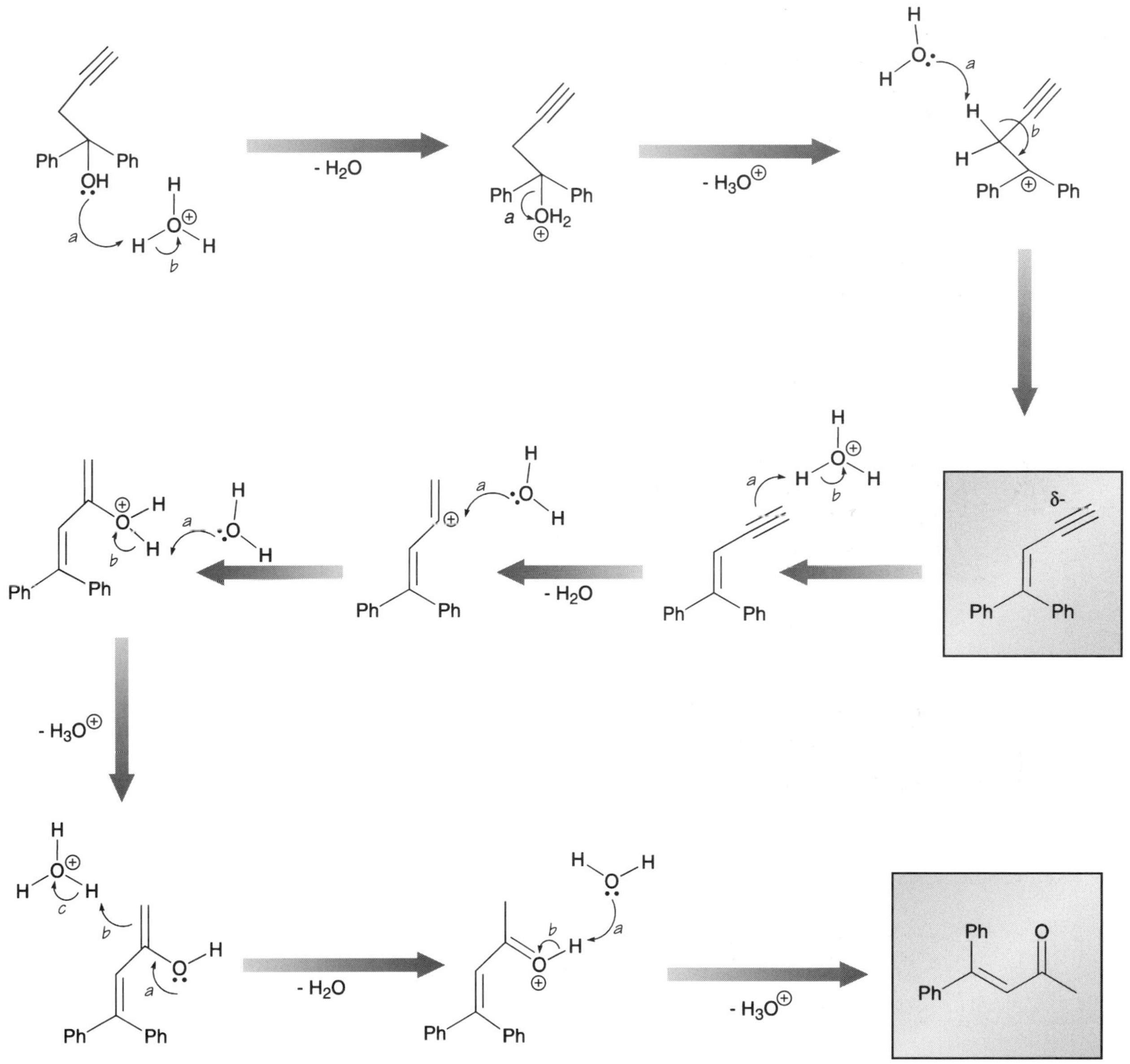

Summary: This is an example of the acid-catalysed hydration of an alkyne.

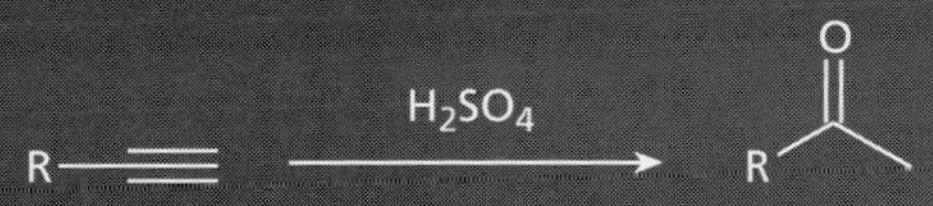

$$R-C\equiv CH \xrightarrow{H_2SO_4} RCOCH_3$$

Now try questions 2.14 and 2.15 at the end of this Chapter

2.5

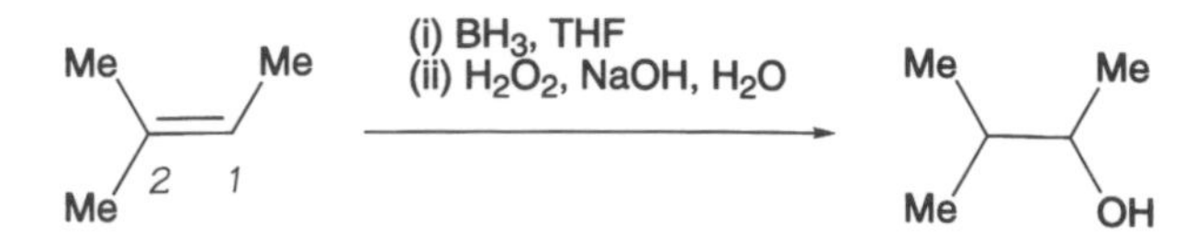

Label electrophilic/nucleophilic and acidic/basic sites of all reactants, and number identical atoms in the starting material and product

Alkenes are nucleophiles by virtue of the π electron density.
Borane is an electrophile (boron is electron-deficient) and is therefore a Lewis acid.

Identify the most reactive sites, if more than one exists

The alkene is the only reactive site.

Recall the characteristic reactions of the most reactive functional groups and, by considering the reaction conditions, decide which is the most appropriate

Hydroboration of alkenes occurs easily, with boron adding to the less hindered carbon.

Work through the mechanism leading to the intermediate product

The **addition** of H and B to the alkene occurs in a syn- process; this is repeated a further two times, using the remaining B–H bonds, to give a **trialkylborane**.

Repeat the above four steps...

The boron atom of the trialkylborane is still electron-deficient, and is readily attacked by hydroperoxide anion (generated from hydrogen peroxide and base). This process induces a *1,2*-**alkyl migration** from B to O, which is repeated a further two times to give a **trialkoxyborane**.

Recognize that this is not the final product, but is closely related to it

The **trialkoxyborane** is easily **hydrolysed** under alkaline conditions to the corresponding alkoxide, which is protonated under the aqueous conditions of the reaction.

Write down the structure of the final product

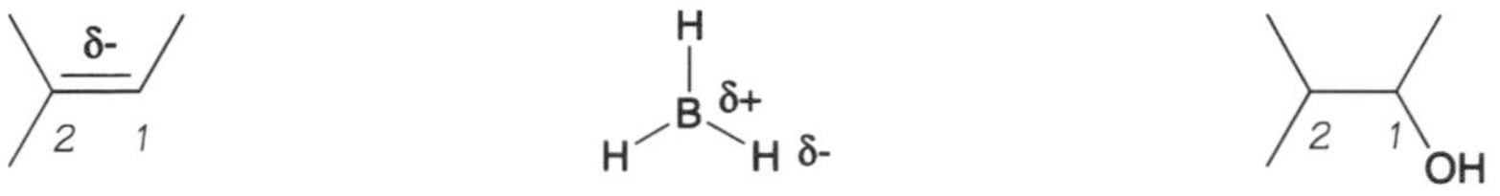

δ-
2 1

H
B δ+
H H δ-

2 1 OH

$$HO^{\ominus} + H{-}OOH \rightleftharpoons H_2O + {}^{\ominus}O{-}OH$$

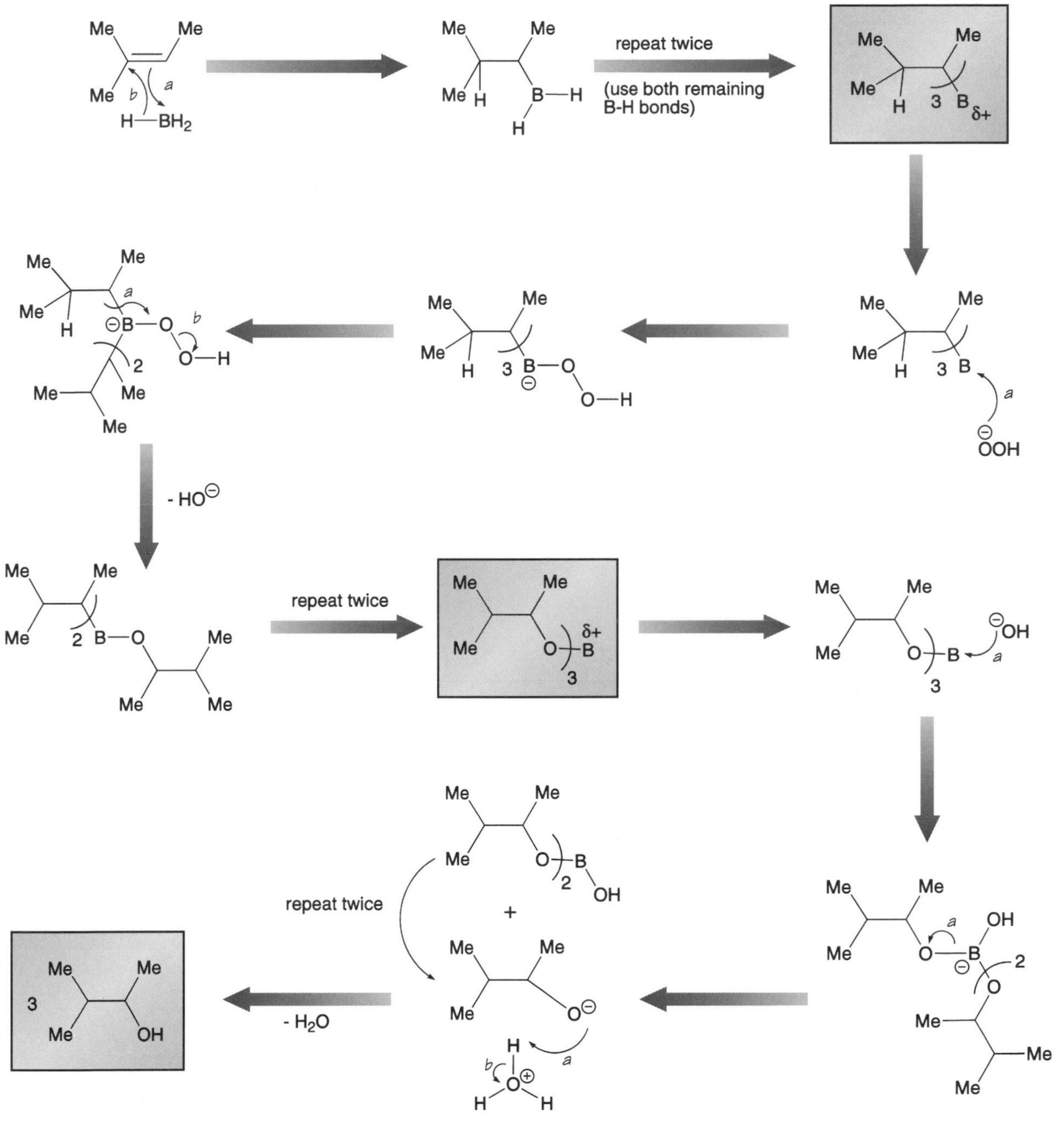

Me Me
Me b a
H—BH₂

Me Me
Me H B—H
H

repeat twice

(use both remaining B-H bonds)

Me Me
Me H 3 B δ+

Me Me
Me H 3 B a
⊖OOH

Me Me
Me H 3 B—O
⊖ O—H

Me Me a
Me H ⊖B—O b
2 O—H
Me— Me
Me

$- HO^{\ominus}$

Me Me
Me 2 B—O Me
Me Me

repeat twice

Me Me
Me O δ+ B
3

Me Me
Me O B ⊖OH a
3

Me Me
Me a OH
O—B
⊖ 2
O
Me—
Me
Me

Me Me
Me O B
2 OH

+

Me Me
Me O⊖

H a
b O⊕
H H

repeat twice

$- H_2O$

Me Me
3
Me OH

Summary: *This is an example of a hydroboration reaction, followed by conversion to the corresponding alcohol.*

$$3RCH{=}CH_2 \xrightarrow{BH_3} (RCH_2CH_2)_3B \xrightarrow[NaOH]{H_2O_2} 3\,RCH_2CH_2OH$$

Now try questions 2.16 and 2.17 at the end of this Chapter

2.6

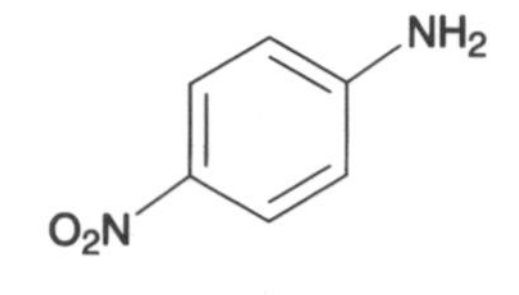

(i) $NaNO_2$, HCl, H_2O

(ii) NaOAc, H_2O,

1 2 3 4

O₂N–C₆H₄–CH₂CH=CHCH₂Cl

Label electrophilic/nucleophilic and acidic/basic sites of all reactants, and number identical atoms in the starting material and product

The amine group is a good nucleophile, although it is deactivated by **resonance** with the nitro group.

Identify the most reactive sites, if more than one exists

Although the phenyl ring is nucleophilic, the amine is the most nucleophilic site.

Recall the characteristic reactions of the most reactive functional groups and, by considering the reaction conditions, decide which is the most appropriate

The combination $NaNO_2$/HCl generates HONO (nitrous acid), which is in **equilibrium** with N_2O_3, a source of $O{=}N^+$. This reagent *diazotizes* the amine functional group.

Work through the mechanism leading to the intermediate product

Nucleophilic addition of the amine is followed by tautomerization, which permits **elimination** of H_2O to give the **diazonium** salt product.

Repeat the above four steps...

The diazo group loses nitrogen gas to give an aryl cation; this is intercepted by the diene to give a stabilized **allylic** cation. The 2° **allylic** cation shown is the more stable.

Recognize that this is not the final product, but is closely related to it

The **allylic** cation reacts with chloride at the least hindered end to give the product.

Write down the structure of the final product

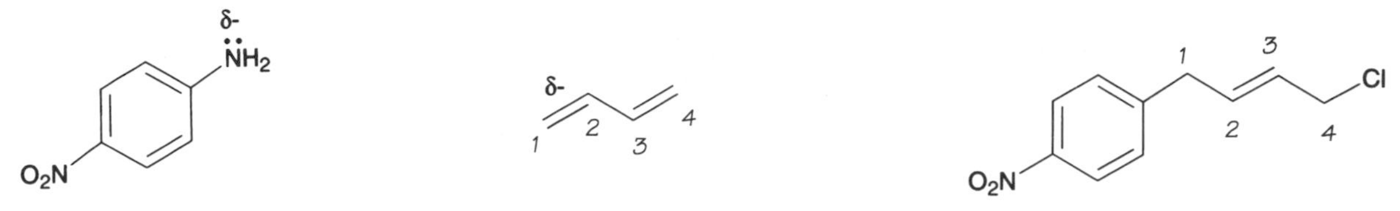

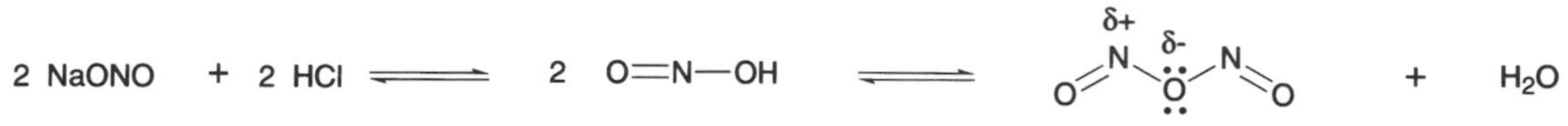

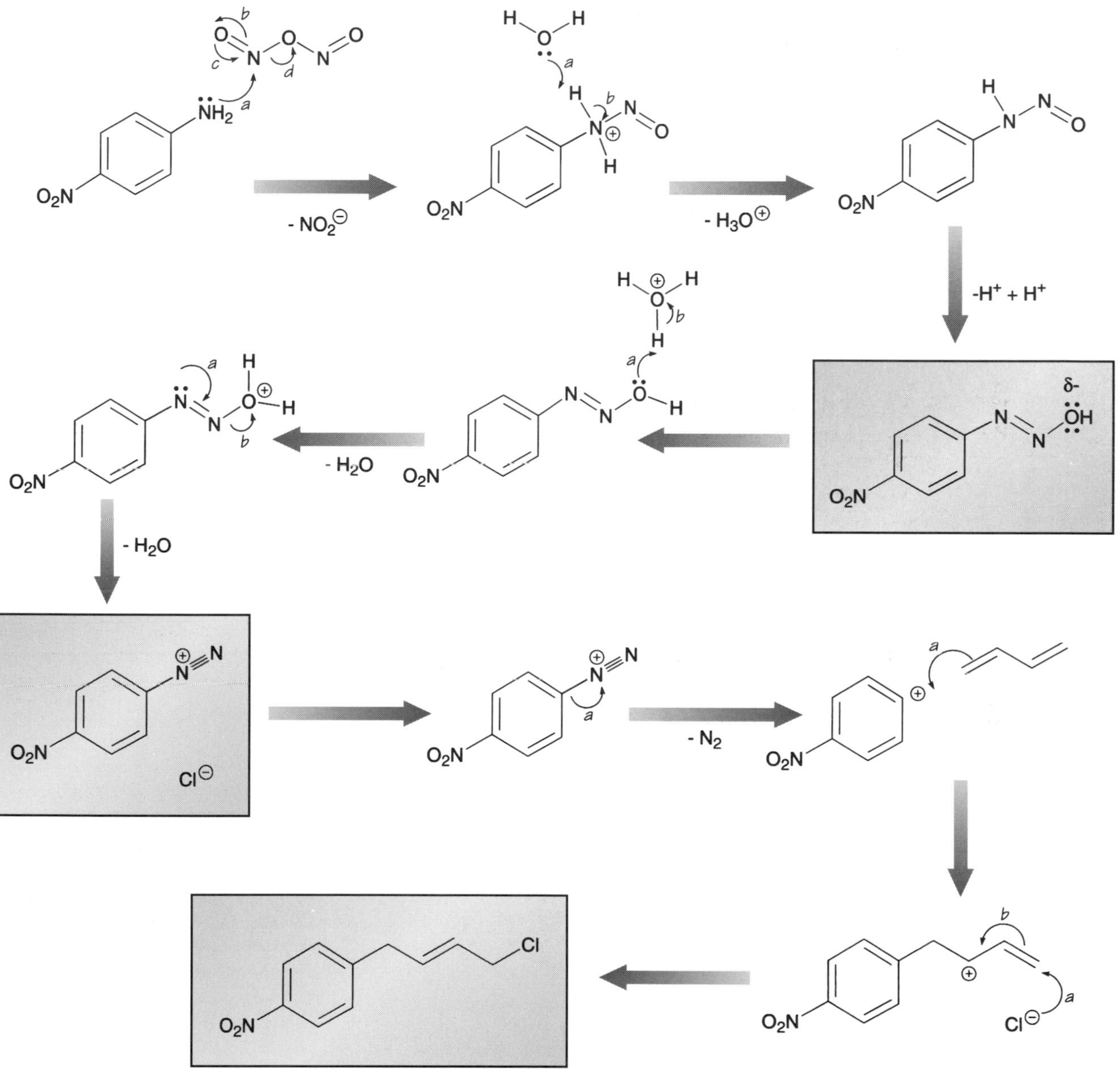

$$2\ NaONO + 2\ HCl \rightleftharpoons 2\ O{=}N{-}OH \rightleftharpoons O{=}N{-}O{-}N{=}O + H_2O$$

- $NO_2^{\ominus}$

- $H_3O^{\oplus}$

$-H^+ + H^+$

- H_2O

- H_2O

- N_2

Summary: This is an example of the reaction of a diazonium ion with nucleophiles.

$$ArN_2^{\oplus} \xrightarrow{Nu^{\ominus}} ArNu$$

Now try questions 2.18 and 2.19 at the end of this Chapter

2.7

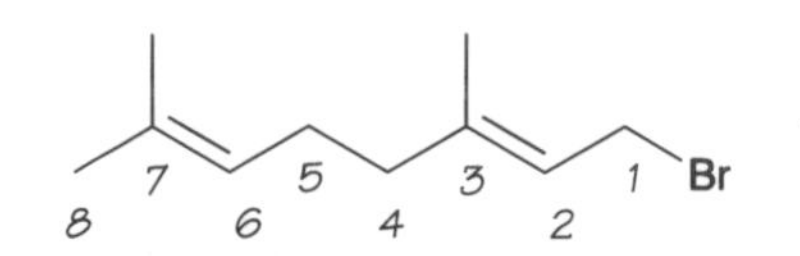

(i) Mg, Et_2O
(ii) ethylene oxide (O), then acidic work-up
(iii) CrO_3, H_2SO_4, H_2O
(iv) H_3PO_4

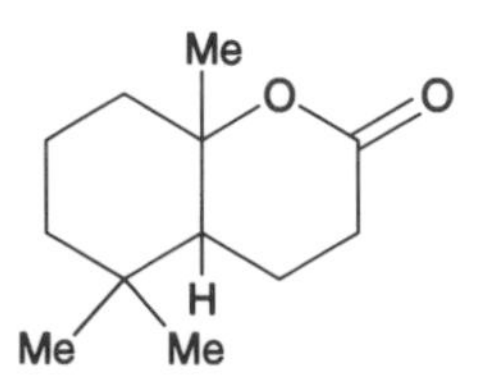

Label electrophilic/nucleophilic and acidic/basic sites of all reactants, and number identical atoms in the starting material and product

Alkyl halides and magnesium react to form Grignard reagents.
Ethylene oxide is an electrophile, with the C–O bond polarized by the electronegative oxygen; there is substantial ring strain present.

Identify the most reactive sites, if more than one exists

The alkyl halide reacts at C-1 to generate a carbanion equivalent. Ethylene oxide is susceptible to nucleophilic attack at the carbon adjacent to the oxygen.

Recall the characteristic reactions of the most reactive functional groups and, by considering the reaction conditions, decide which is the most appropriate

Grignard reagents react with ethylene oxide to give alcohols in which the alkyl chain has been extended by two carbon atoms.

Work through the mechanism leading to the intermediate product

Nucleophilic attack (S_N2) by the Grignard at ethylene oxide gives an alkoxide; protonation on acid work-up gives an alcohol.

Repeat the above four steps...

* The alcohol undergoes nucleophilic addition to chromium trioxide, to give a chromate ester which collapses to the aldehyde; this aldehyde is in **equilibrium** with its *hydrate*, and the oxidation process with chromium trioxide is repeated, to give a carboxylic acid.
* Phosphoric acid, a strong acid, protonates the terminal double bond (**Markovnikov addition**).

Recognize that this is not the final product, but is closely related to it

The carboxylic acid function acts as an internal nucleophile, to give the bicyclic lactone product.

Write down the structure of the final product

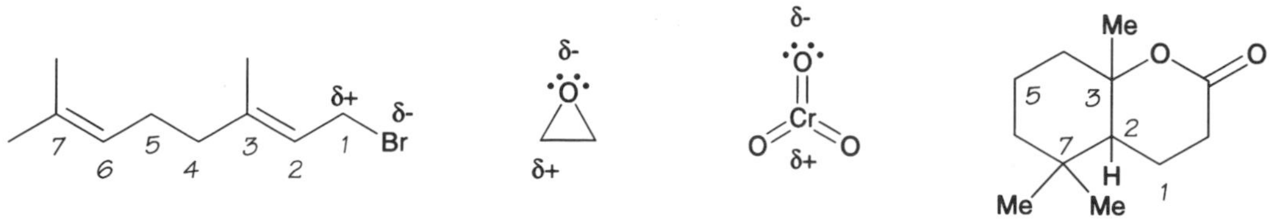

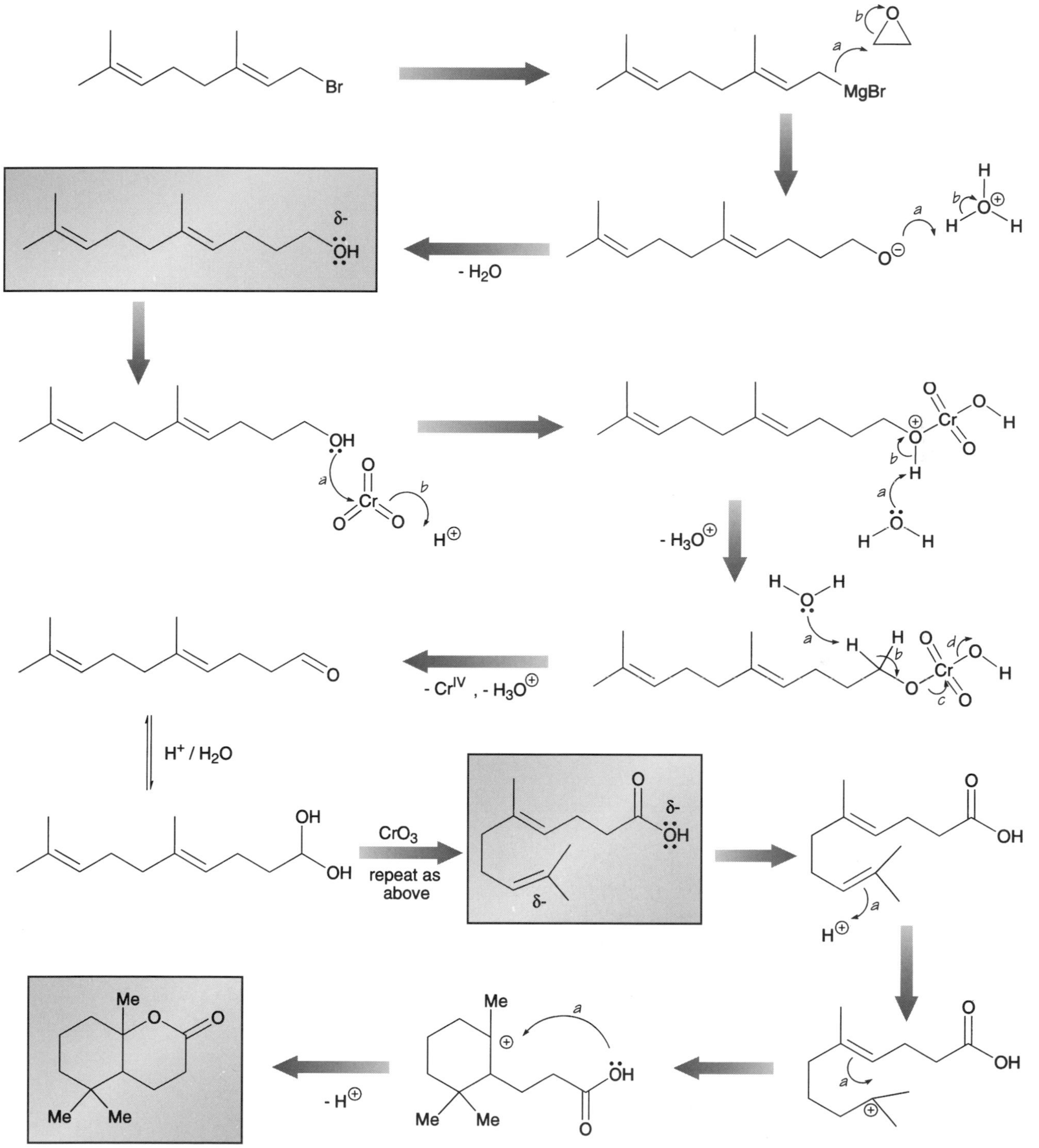

Summary: This is an example of intramolecular trapping of an alkyl cation with a carboxylic acid to give a lactone.

Now try questions 2.20 and 2.21 at the end of this Chapter

Supplementary questions

2.8

(i) H_2SO_4
(ii) H_2, Pd/C

2.9

(i) Br_2, H_2O
(ii) NaOH, H_2O

2.10

HOCl, HOAc, H_2O

(±)

2.11

(i) $SOCl_2$, py
(ii) NaOEt, EtOH

2.12

HCO_3H

2.13

(i) I_2, AcOAg
(ii) AcOH, H_2O
(iii) KOH, H_2O

2.14

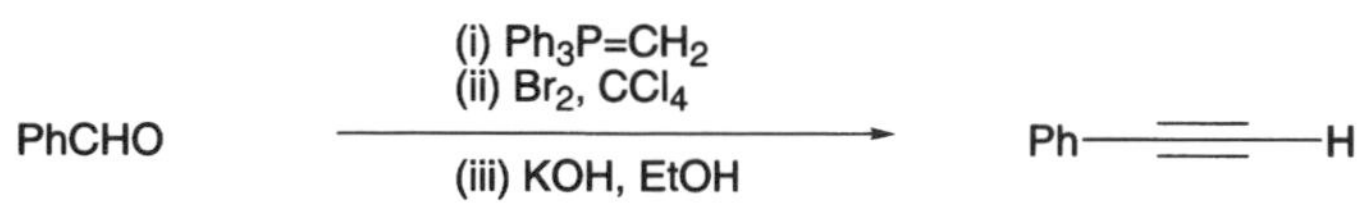

PhCHO —(i) $Ph_3P{=}CH_2$; (ii) Br_2, CCl_4; (iii) KOH, EtOH→ Ph—≡—H

2.15

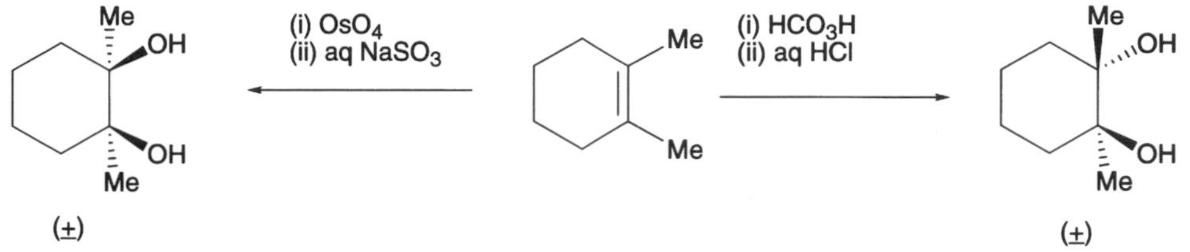

(±) ← (i) OsO_4; (ii) aq $NaSO_3$ — [1,2-dimethylcyclohexene] — (i) HCO_3H; (ii) aq HCl → (±)

Me, OH, OH, Me; Me, Me; Me, OH, OH, Me

2.16

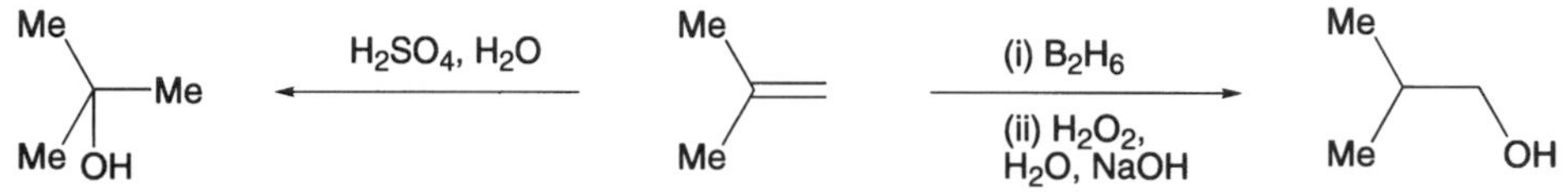

Me, Me, Me, OH ← H_2SO_4, H_2O — Me, Me — (i) B_2H_6; (ii) H_2O_2, H_2O, NaOH → Me, Me, OH

2.17

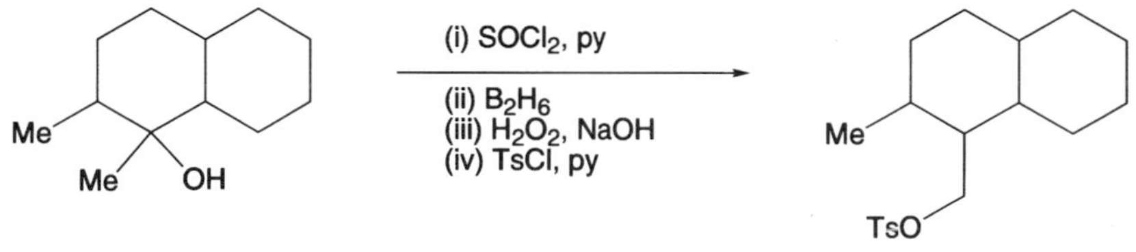

Me, Me, OH —(i) $SOCl_2$, py; (ii) B_2H_6; (iii) H_2O_2, NaOH; (iv) TsCl, py→ Me, TsO

2.18

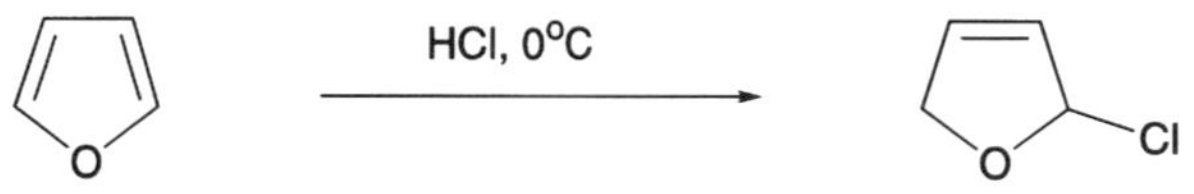

O —HCl, 0°C→ O, Cl

2.19

O —Br_2, MeOH, Na_2CO_3→ MeO, O, OMe

2.20

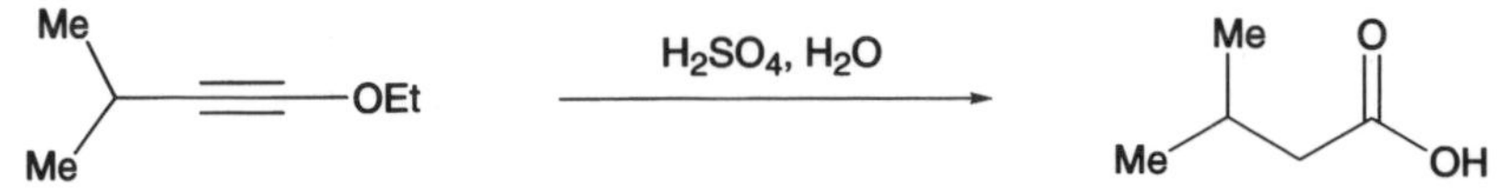

H_2SO_4, H_2O

2.21

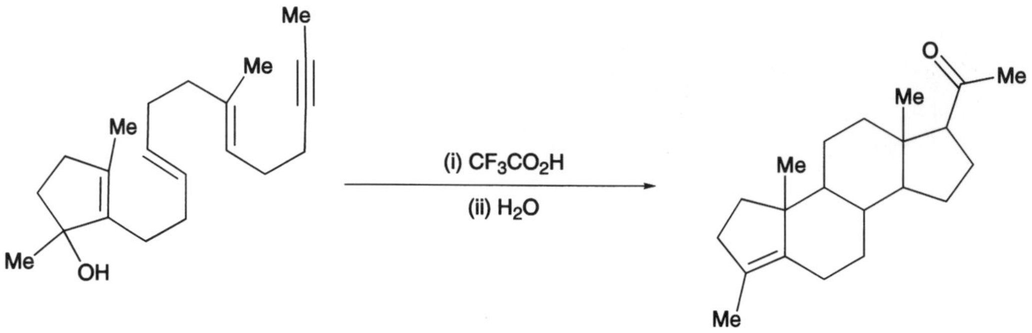

(i) CF_3CO_2H

(ii) H_2O

3 Nucleophilic additions to carbonyl groups

Nucleophilic addition is a typical reaction of all carbonyl groups because of the polarization of the C–O bond, caused by the electronegativity difference of the carbon and oxygen atoms and the presence of a π bond:

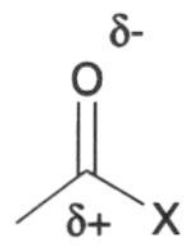

Typical reactions depend on the nature of X and are therefore:

Aldehydes and ketones (X = H, R, respectively)

1 Irreversible addition of nucleophiles
- (a) Reduction. Addition of hydrogen, with:
  - H_2 and a catalyst.
  - 'H^-' from various hydride ion sources, such as complex metal hydrides ($LiAlH_4$ and $NaHB_4$).
  - 'Organic' hydride sources, such as the Cannizzaro reaction and the Meerwein–Ponndorf–Verley reaction.
- (b) Carbanion addition. Addition of Grignard reagents, organolithium reagents and acetylide anions.

2 Reversible addition of nucleophiles

There are three types of carbonyl additions, depending on the nature of the nucleophile:
- (a) Addition. Bisulphite, cyanide.
- (b) Addition–substitution. This mechanism is important for group 6 nucleophiles (i.e. oxygen and sulfur), leading to their respective products (hemiacetals and acetals, and thioacetals).
- (c) Addition–elimination. This mechanism is important for group 5 nucleophiles (i.e. ammonia, hydroxylamine, hydrazine, semicarbazide, 2,4-dinitrophenylhydrazine), to give the products (imines, oximes, hydrazones, semicarbazones, 2,4-dinitrophenylhydrazones, respectively). The Wittig reaction is a special case using a phosphorus ylide as the nucleophile.

Acyl compounds (X = OH (acids), X = OR (esters), X = NR_2 (amides), X = Cl, Br (acid halides) and X = OC(O)R (acid anhydrides))

Unlike aldehydes and ketones, in these acyl derivatives X is a leaving group (a good leaving group is the conjugate base of a strong acid), and so the typical reactivity of any carbonyl group, that of nucleophilic addition, is almost invariably followed by elimination. The overall process is therefore an addition–elimination reaction, *via* a tetrahedral intermediate, which regenerates the carbonyl group after expulsion of the leaving group X^-.

The reactivity order for nucleophilic additions to carbonyl groups is:

$$RCOCl,\ RC(O)O(O)R > RCHO > R_2CO > RCO_2R' > RCONR'_2$$

The order of reactivity is dictated by the steric and electronic effects of the substituent X.

3.1

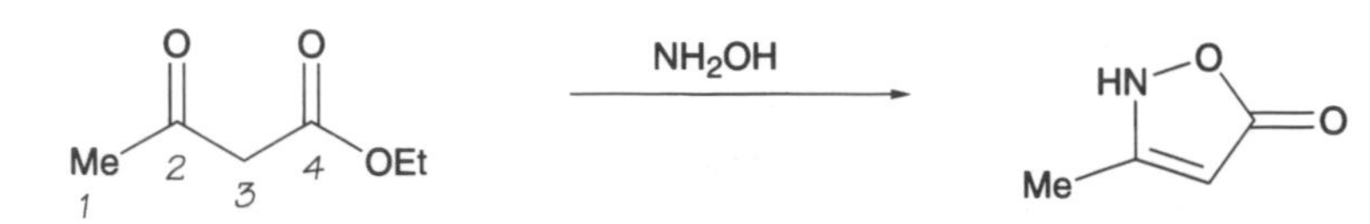

Label electrophilic/nucleophilic and acidic/basic sites of all reactants, and number identical atoms in the starting material and product

The carbonyl group is polarized by the **electronegativity** difference between carbon and oxygen, making the carbon electrophilic.
Hydroxylamine is nucleophilic, since both nitrogen and oxygen atoms possess lone pairs.

Identify the most reactive sites, if more than one exists

Ketones are more reactive than esters, due to the stabilizing resonance delocalization in the ester, and the nitrogen of hydroxylamine is more nucleophilic than oxygen, since nitrogen is less electronegative.

Recall the characteristic reactions of the most reactive functional groups and, by considering the reaction conditions, decide which is the most appropriate

Hydroxylamine and ketones react to form **oximes**.

Work through the mechanism leading to the intermediate product

Oxime formation with hydroxylamine in a standard **addition–elimination** mechanism at a ketone.

Repeat the above four steps...

The electrophilic ester carbonyl is susceptible to **nucleophilic addition**; the most nucleophilic site now available is the oxygen of the oxime. Alcohols and esters react to give esters, and here there is lactone formation by a standard **addition–elimination** mechanism at an ester.

Recognize that this is not the final product, but is closely related to it

The intermediate product can be converted to the final product by **tautomerization**. This can be catalysed by any available acid or base.

Write down the structure of the final product

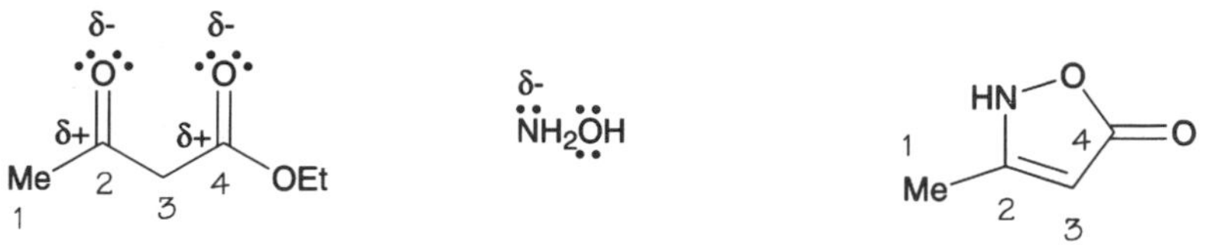

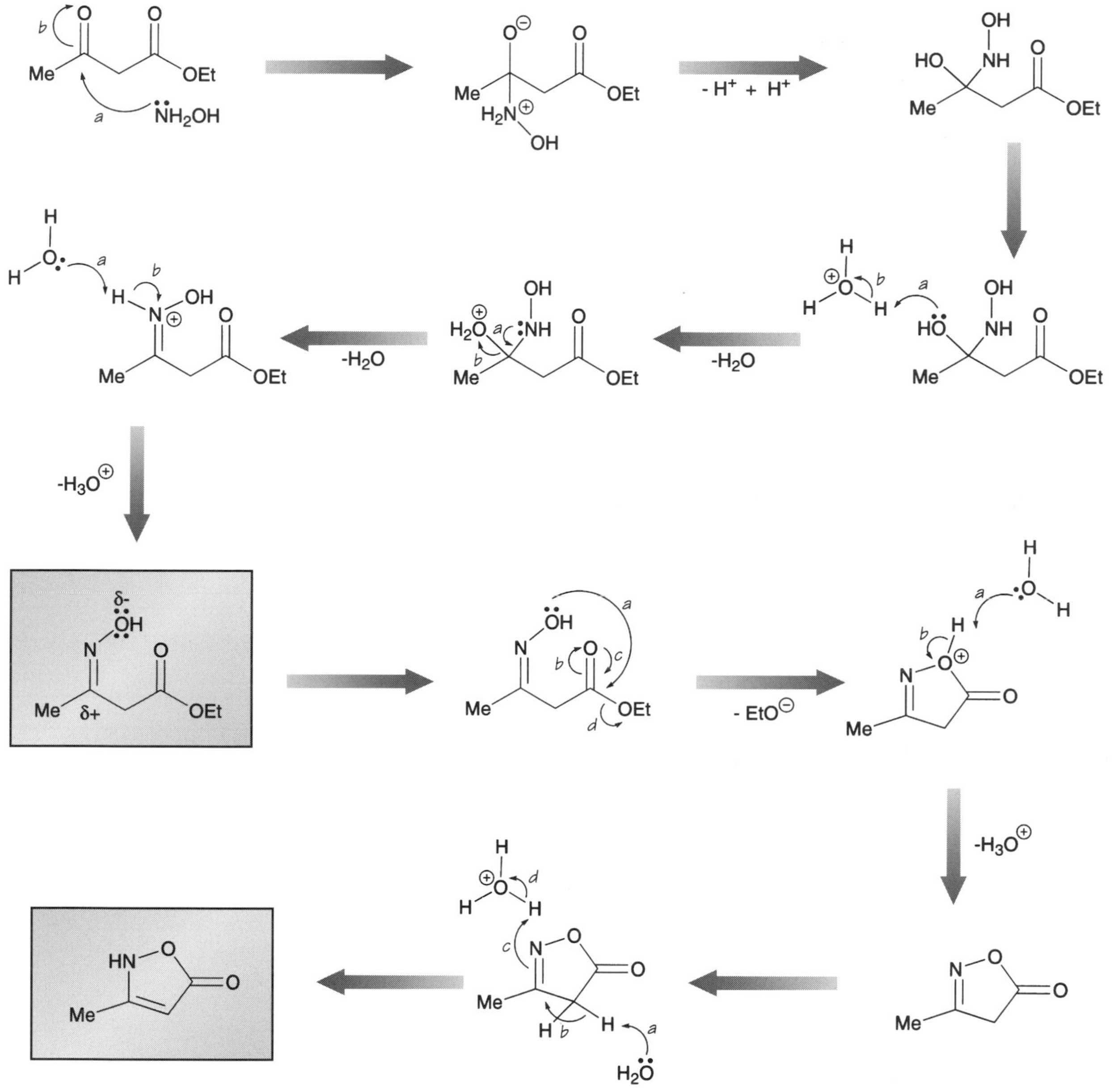

Summary: Carbonyl compounds readily undergo addition–elimination reactions.

Now try questions 3.8 and 3.9 at the end of this Chapter

3.2

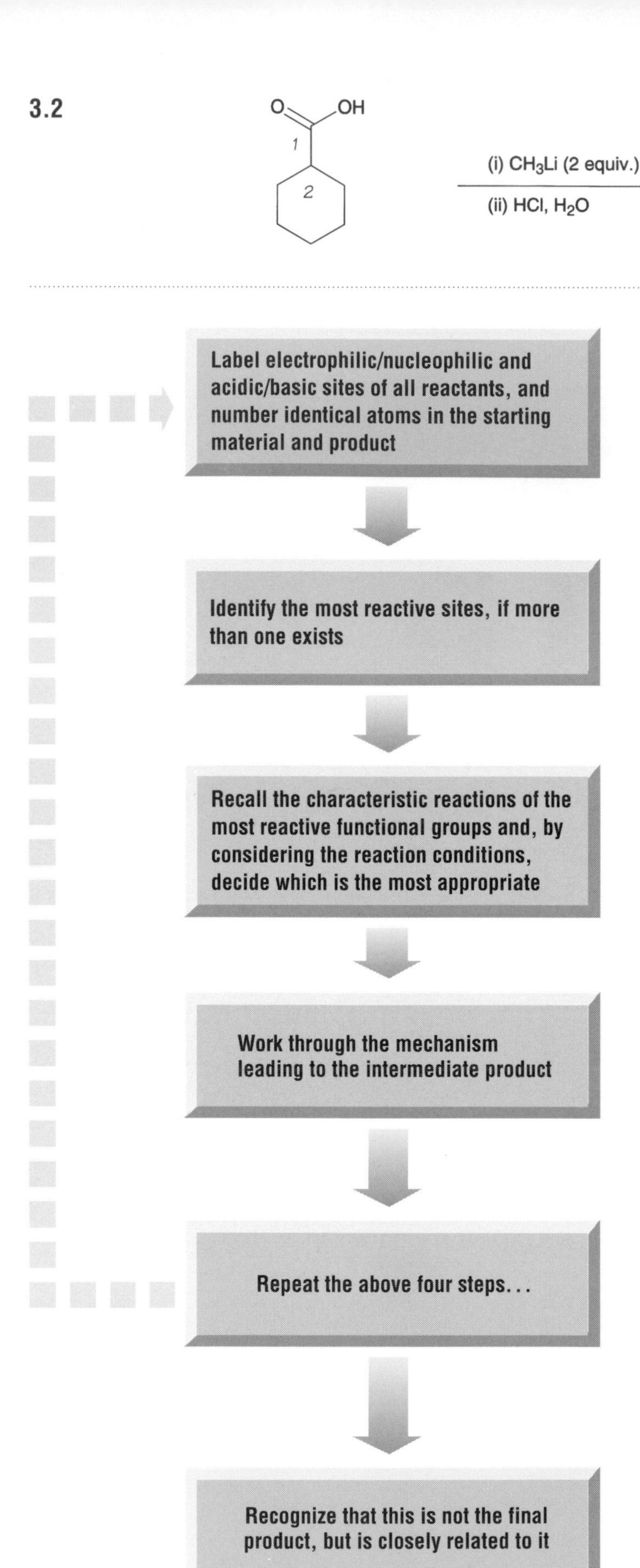

O
OH
1
2
Label electrophilic/nucleophilic and acidic/basic sites of all reactants, and number identical atoms in the starting material and product
Identify the most reactive sites, if more than one exists
Recall the characteristic reactions of the most reactive functional groups and, by considering the reaction conditions, decide which is the most appropriate
Work through the mechanism leading to the intermediate product
Repeat the above four steps...
Recognize that this is not the final product, but is closely related to it
Write down the structure of the final product

(i) CH_3Li (2 equiv.)

(ii) HCl, H_2O

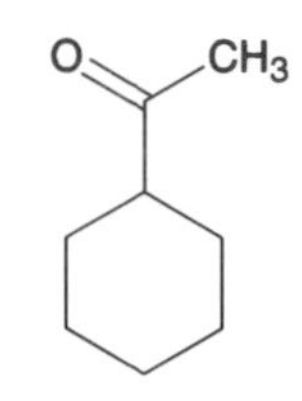

The hydroxylic hydrogen of a carboxylic acid is especially acidic (resonance-stabilized anion).
Methyl lithium is a carbanion equivalent, and is a powerful base and nucleophile.

The carboxyl group is the only reactive electrophilic site here.

Acid–base proton exchange will always occur in preference to nucleophilic attack if both are possible; here the carboxylic acid is initially deprotonated by MeLi.
MeLi is a sufficiently powerful nucleophile for it to be able to undergo **nucleophilic addition** to the carboxylate anion, even though it is already negatively charged.

Nucleophilic addition of Me⁻ to the carbonyl group generates a stable tetrahedral intermediate, which gives a ketal after acid treatment.

The product is a ketal, which will undergo dehydration, by **elimination** of water under aqueous acidic conditions; the loss of water is assisted by one of the lone pairs of the ketal oxygens.

Deprotonation of the **oxonium** ion gives the product.

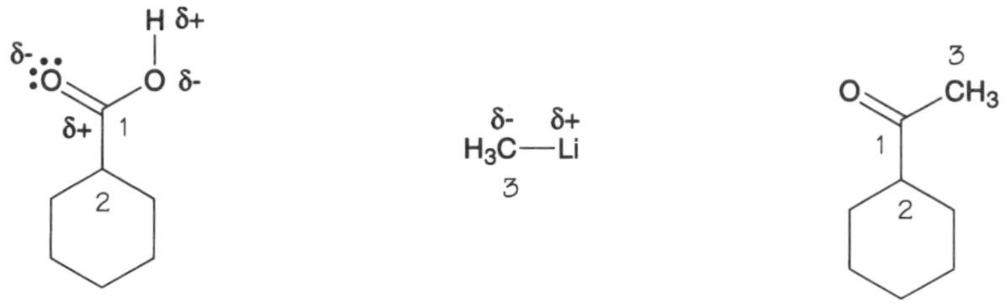

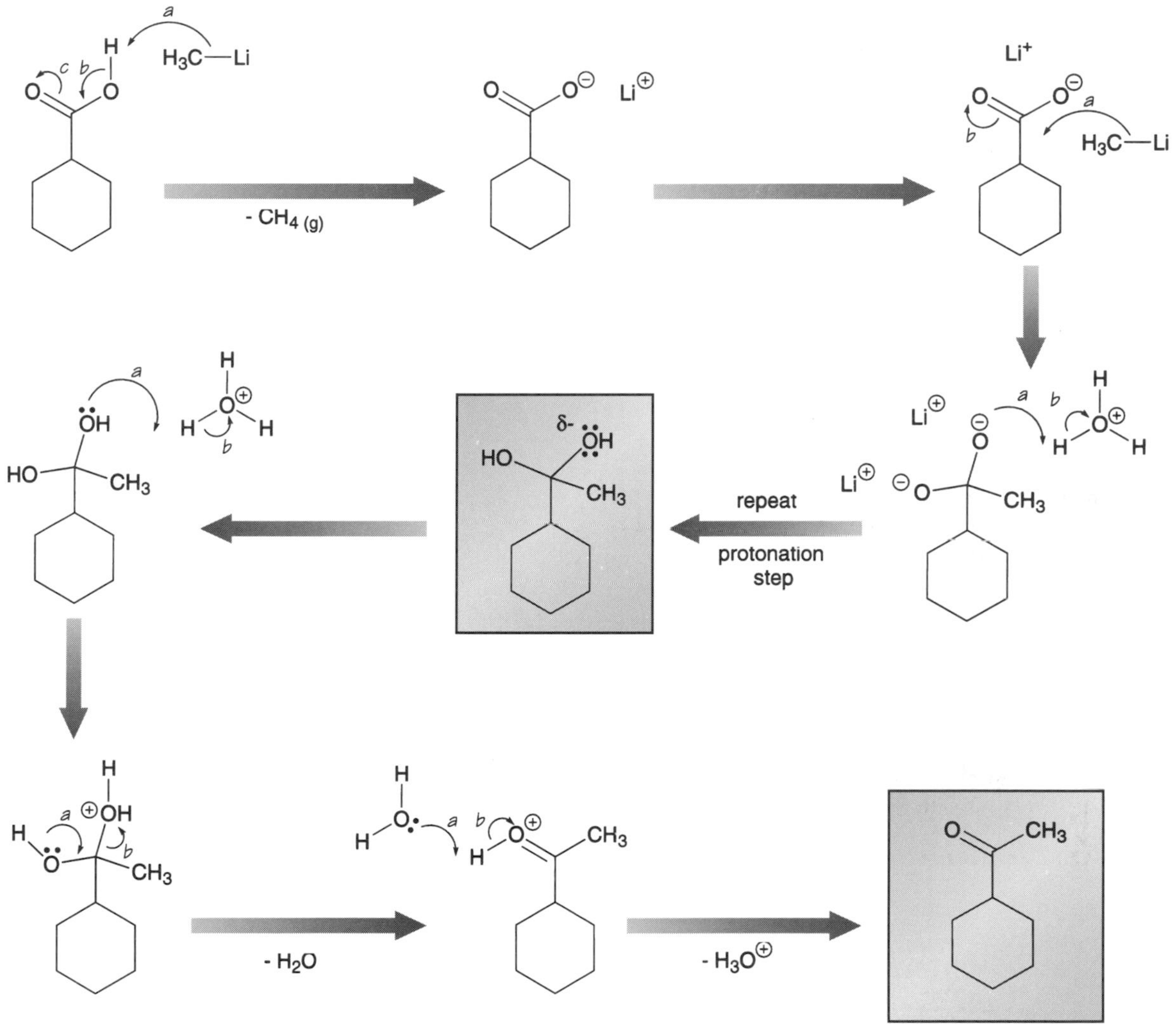

H δ+
δ- :O: O δ-
δ+ 1
2

δ- δ+
H_3C—Li
3

3
O CH_3
1
2

---

a
H H_3C—Li
c b
O O

- $CH_{4\,(g)}$

O $O^{\ominus}$ $Li^{\oplus}$

Li^{+}
O $O^{\ominus}$ a
b
H_3C—Li

$Li^{\oplus}$ a b H
$\ominus$O H $O^{\oplus}$
H H
$Li^{\oplus}$ $^{\ominus}$O CH_3

repeat
protonation
step

δ- ..
HO ÖH
CH_3

a H
ÖH H $O^{\oplus}$ H
b
HO CH_3

H
a $^{\oplus}$OH
H b
Ö CH_3

- H_2O

H
H O: a b $O^{\oplus}$ CH_3
H

- $H_3O^{\oplus}$

O CH_3

Summary: *Carboxylic acids react with 2 equivalents of an organolithium reagent to give a ketone.*

$$RCO_2H \quad + \quad 2R'Li \xrightarrow{\text{acidic work-up}} RC(=O)R'$$

Now try questions 3.10 and 3.11 at the end of this Chapter

3.3

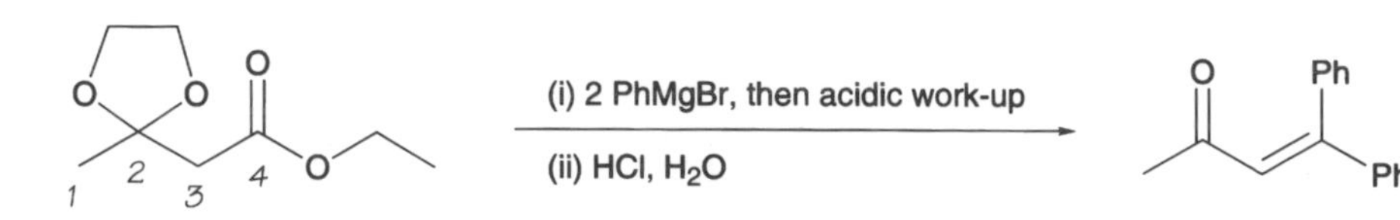

Label electrophilic/nucleophilic and acidic/basic sites of all reactants, and number identical atoms in the starting material and product

The carbonyl group is polarized by the **electronegativity** difference between carbon and oxygen, making the carbon electrophilic.
Phenylmagnesium bromide is a *carbanion equivalent*.

Identify the most reactive sites, if more than one exists

The ester function is the only reactive site here (it is an electrophile), as an acetal is inert to all conditions except acidic **hydrolysis**.

Recall the characteristic reactions of the most reactive functional groups and, by considering the reaction conditions, decide which is the most appropriate

Nucleophilic *addition–elimination* to an ester carbonyl, with alkoxide as the leaving group.

Work through the mechanism leading to the intermediate product

Addition–elimination of 1 equiv. of PhMgBr gives a phenyl ketone. This is still electrophilic (carbonyl polarized by electronegativity difference between C and O). A second addition of PhMgBr then occurs, which after protonation on work-up gives a **tertiary alcohol**.

Repeat the above four steps...

The acetal is readily **hydrolysed** by aqueous acid, by the usual protonation–elimination–addition of water–elimination sequence for acetal hydrolysis.

Recognize that this is not the final product, but is closely related to it

Tertiary alcohols are easily dehydrated under acidic conditions by **elimination** of water.

Write down the structure of the final product

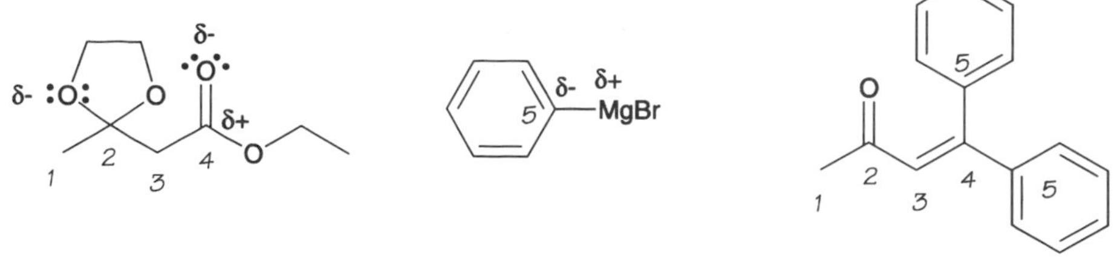

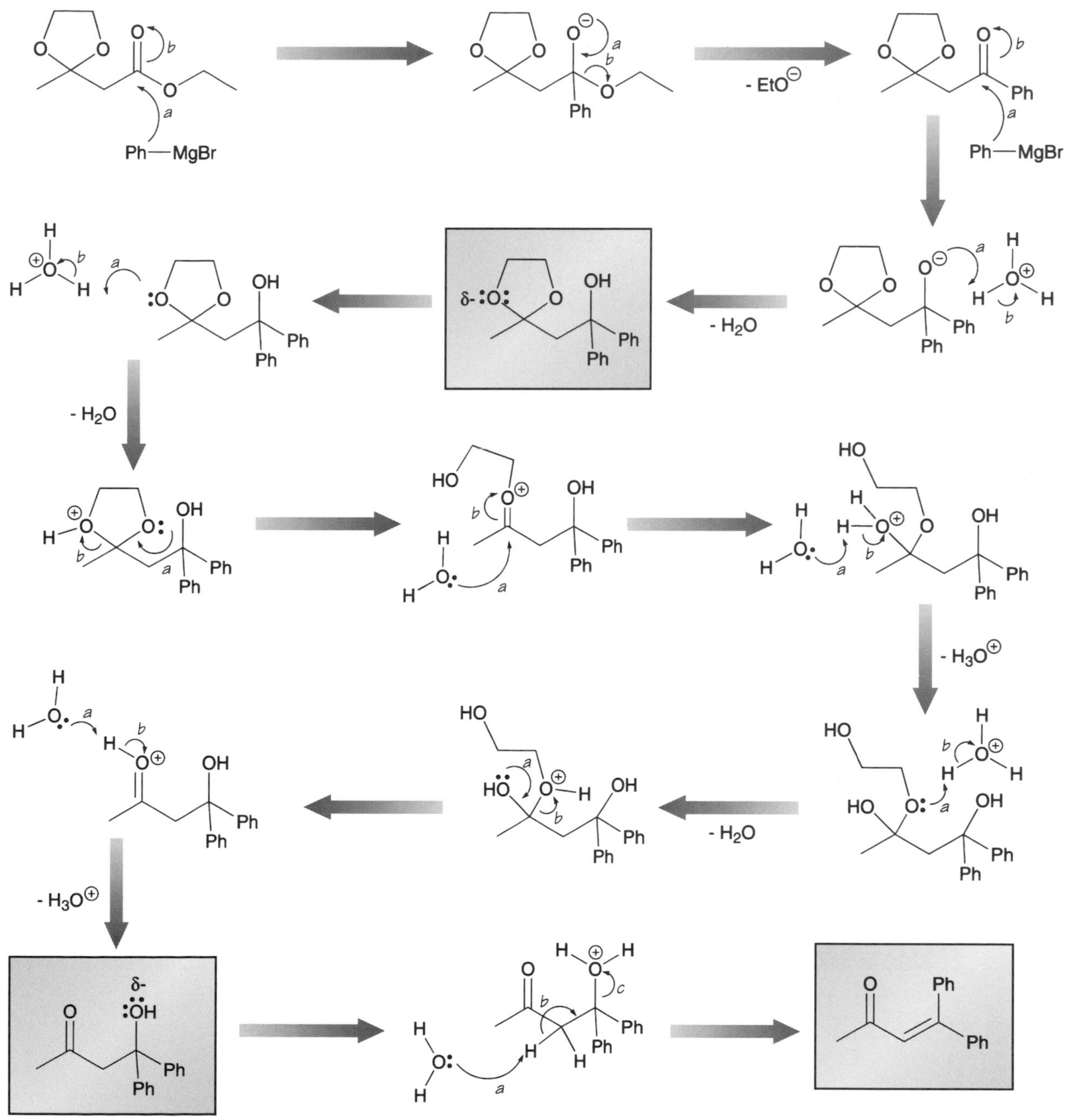

Summary: An ester reacts with 2 equivalents of Grignard to give a tertiary alcohol.

$$RCO_2R' + 2R''MgBr \longrightarrow R\text{–}C(OH)(R'')\text{–}R''$$

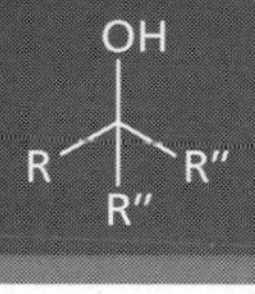

Now try questions 3.12 and 3.13 at the end of this Chapter

3.4

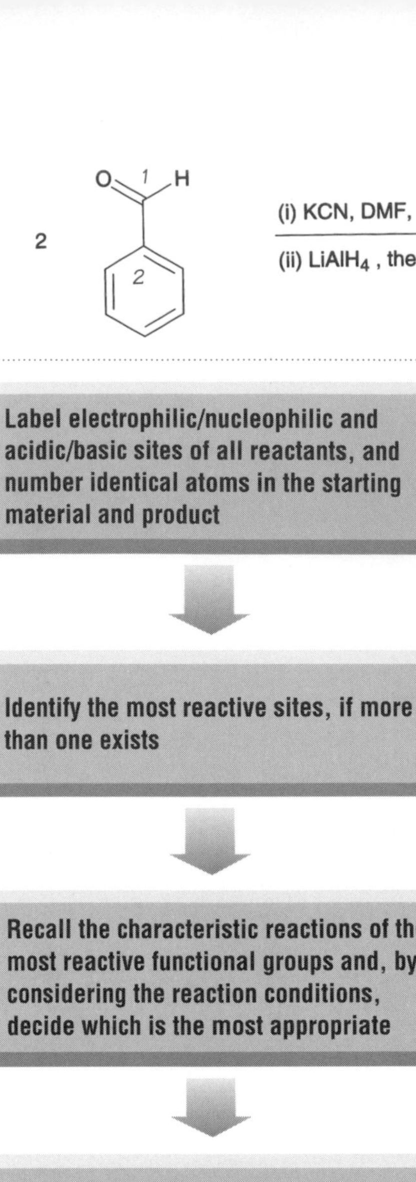

(i) KCN, DMF, H_2O

(ii) $LiAlH_4$, then acidic work-up

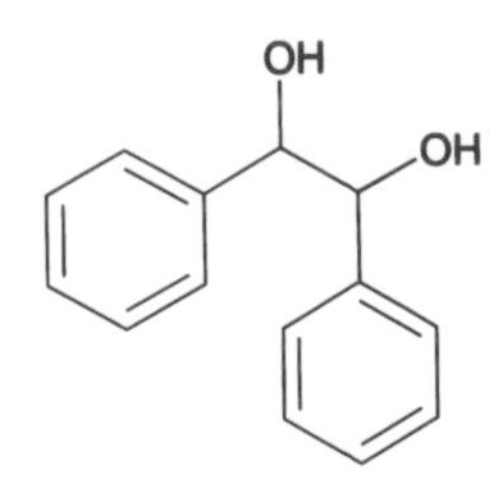

Label electrophilic/nucleophilic and acidic/basic sites of all reactants, and number identical atoms in the starting material and product

The carbonyl group is polarized by the **electronegativity** difference between carbon and oxygen, making the carbon electrophilic.
Cyanide is a good nucleophile and weak base.

Identify the most reactive sites, if more than one exists

The aldehyde is the only electrophilic site.

Recall the characteristic reactions of the most reactive functional groups and, by considering the reaction conditions, decide which is the most appropriate

Reversible addition of NC^- to a carbonyl gives the cyanohydrin product; this has a very acidic α-hydrogen (due to an inductively and resonance-stabilized anion) and this is easily removed by base.

Work through the mechanism leading to the intermediate product

This generates a nucleophile which is able to undergo **nucleophilic addition** to another molecule of benzaldehyde.
After proton transfers, this intermediate collapses to a ketone (benzoin).

Repeat the above four steps...

The ketone product is readily reduced by $LiAlH_4$, by nucleophilic addition of hydride anion (H^-).

Recognize that this is not the final product, but is closely related to it

Acidic work-up gives the 1,2-diol product (hydrobenzoin).

Write down the structure of the final product

$K^{\oplus}$ $^{\ominus}C\equiv N$

$LiAlH_4$ = "$H^{\ominus}$"

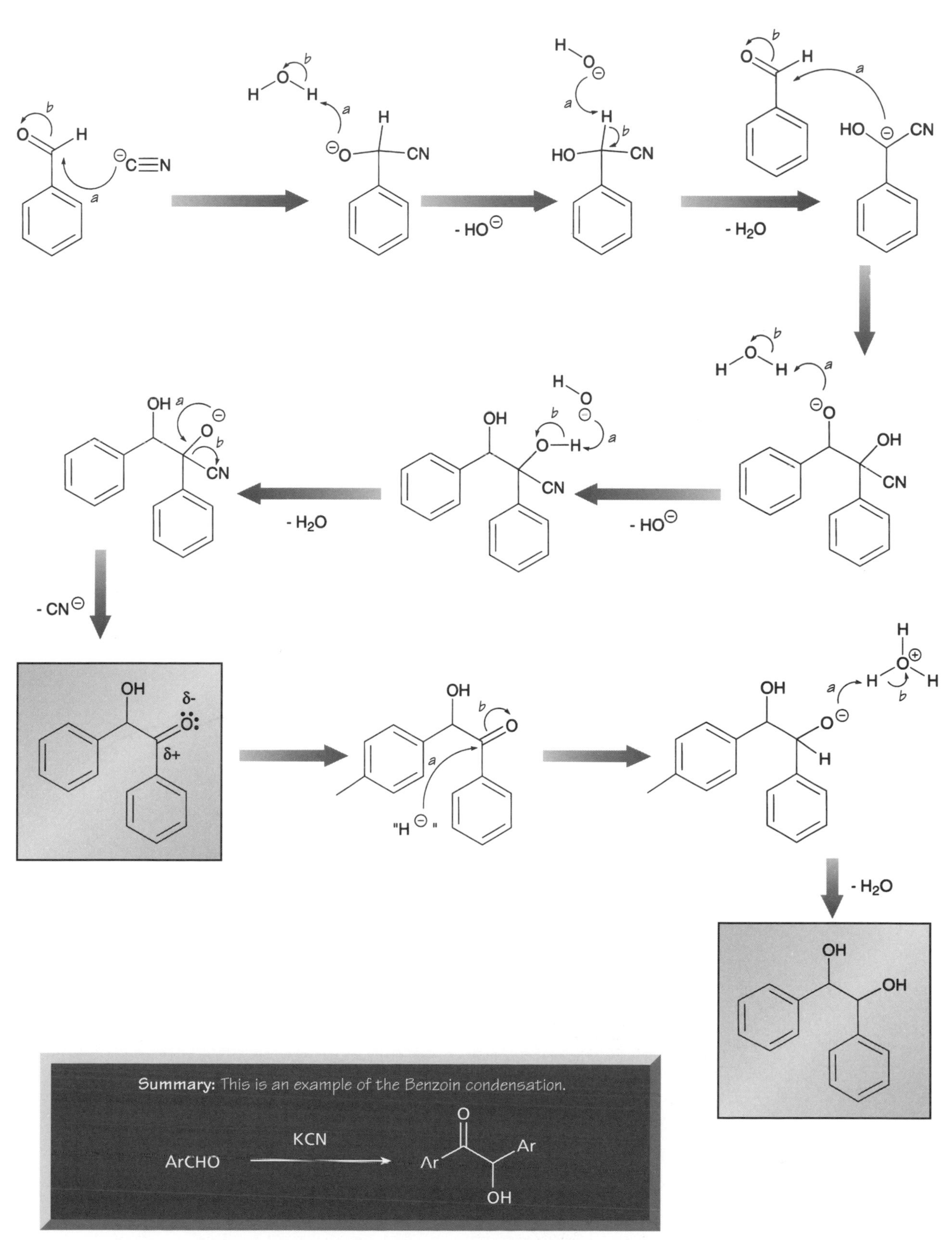

Now try questions 3.14 and 3.15 at the end of this Chapter

3.5

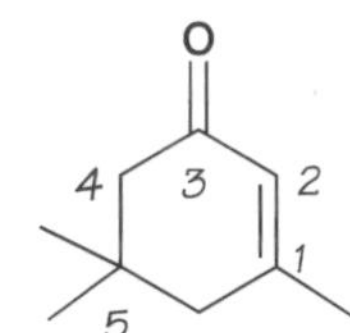

(i) HOOH, NaOH
(ii) $TsNHNH_2$, HOAc
(iii) NaOH

Label electrophilic/nucleophilic and acidic/basic sites of all reactants, and number identical atoms in the starting material and product

α,β-Unsaturated ketones are electrophilic at both the carbon of the carbonyl group and at the β position (here labelled C-1), by virtue of the **electronegativity** difference of the carbon and oxygen atoms.
HOOH is a good nucleophile (both oxygens possess lone pairs) and has a weak O–O bond.

Identify the most reactive sites, if more than one exists

Nucleophiles preferentially add to the β position of an α,β-unsaturated ketone. NaOH and HOOH set up an *equilibrium*, which generates hydroperoxide anion, which is highly nucleophilic.

Recall the characteristic reactions of the most reactive functional groups and, by considering the reaction conditions, decide which is the most appropriate

Conjugate addition of nucleophiles to an α,β-unsaturated ketone.

Work through the mechanism leading to the intermediate product

Conjugate addition of HOO^- gives an enolate anion, which is also a good nucleophile. Since the hydroperoxide itself carries a good leaving group (OH), internal attack to give an epoxide then occurs.

Repeat the above four steps...

* The product epoxy ketone has an electrophilic carbonyl group, and is susceptible to attack by $TsNHNH_2$ (a good nucleophile). Attack of tosylhydrazine gives the corresponding hydrazone by an **addition–elimination** mechanism.
* The base HO^- is easily able to abstract the acidic N–H. Opening of the epoxide and fragmentation of this intermediate gives the observed product directly.

Recognize that this is not the final product, but is closely related to it

Not needed here.

Write down the structure of the final product

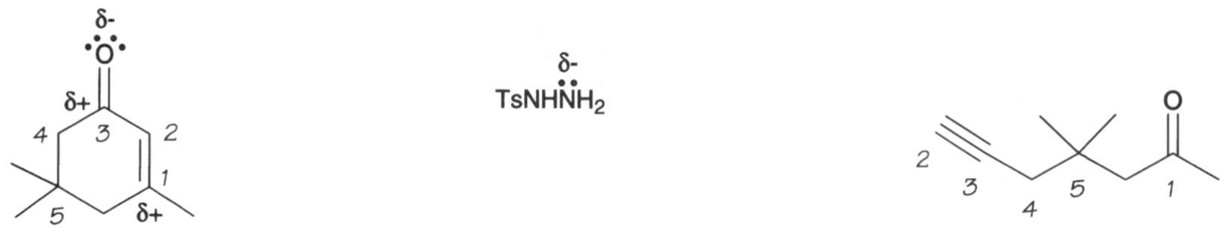

$$HOO{-}H + {}^{\ominus}OH \rightleftharpoons HOO^{\ominus} + H_2O$$

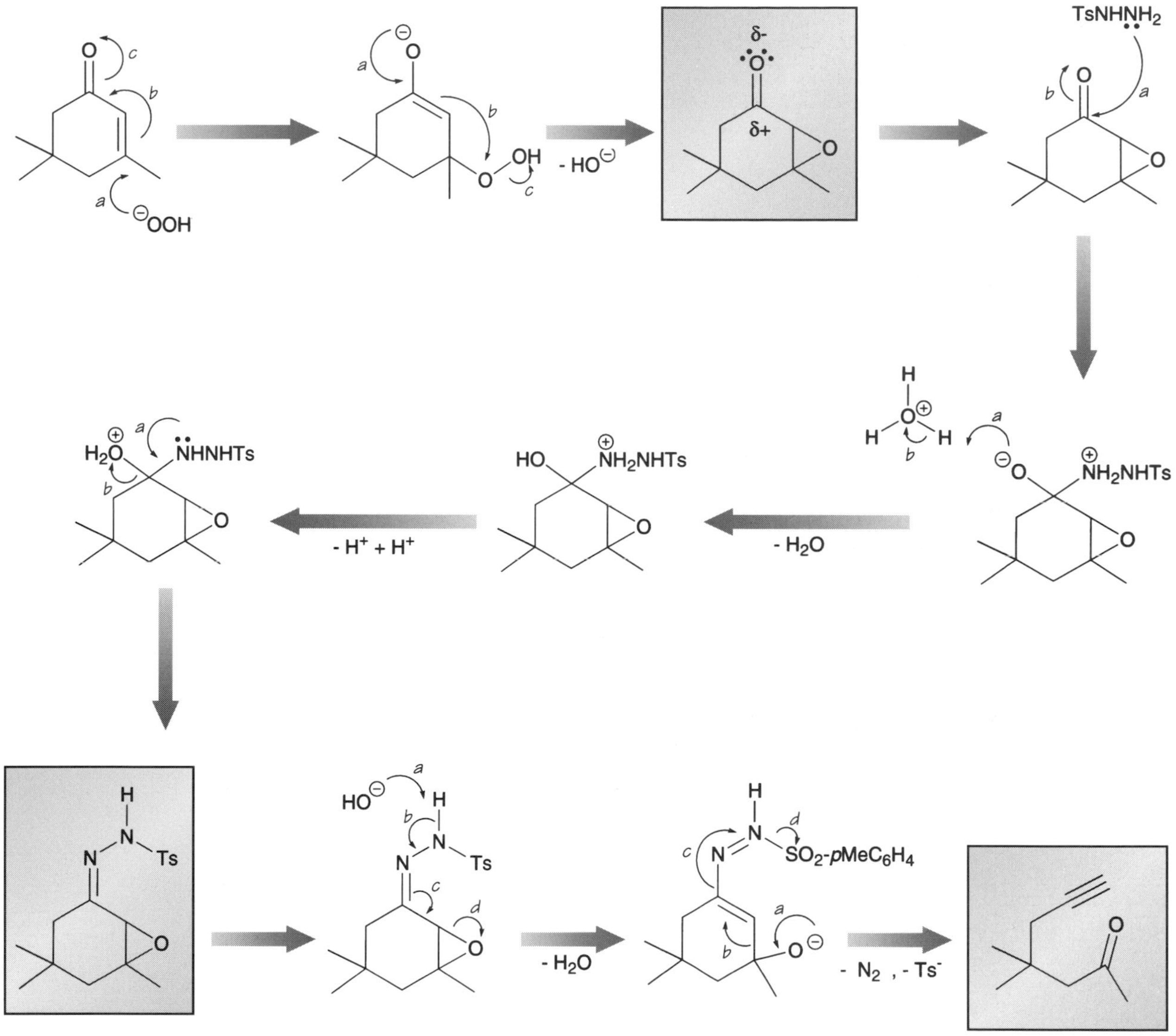

- H$_2$O

- H$^+$ + H$^+$

- H$_2$O

- N$_2$, - Ts$^-$

Summary: Addition of nucleophiles to an α, β-unsaturated enone preferentially occurs at the β-position. This process is called conjugate addition.

Now try questions 3.16 and 3.17 at the end of this Chapter

3.6

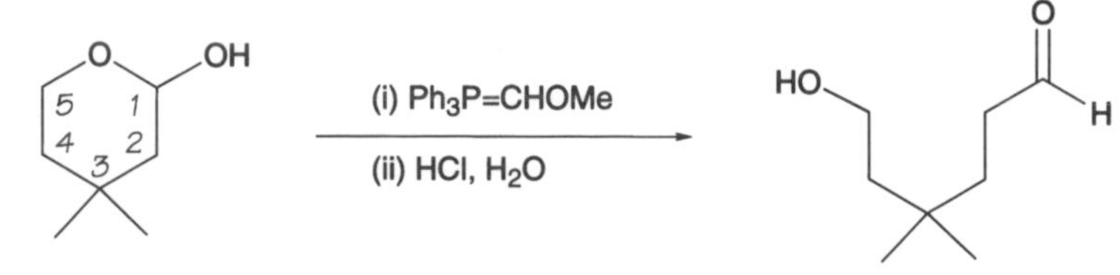

Label electrophilic/nucleophilic and acidic/basic sites of all reactants, and number identical atoms in the starting material and product

The starting material is a cyclic **hemiacetal**, which is in **equilibrium** with the corresponding hydroxy aldehyde, which is as usual electrophilic at the carbonyl carbon.
Ylides are nucleophilic at the carbon adjacent to the P atom; this can be seen from the alternative resonance structure.

Identify the most reactive sites, if more than one exists

The aldehyde is the only electrophilic site.

Recall the characteristic reactions of the most reactive functional groups and, by considering the reaction conditions, decide which is the most appropriate

Addition of the nucleophilic ylide to the carbonyl group then occurs in the usual **Wittig** reaction.

Work through the mechanism leading to the intermediate product

Reversible addition of the ylide is followed by **oxaphosphetane** formation. **Elimination** of $Ph_3P{=}O$ (very stable P=O bond) then gives the *enol ether* product.

Repeat the above four steps...

Enol ethers are both very basic and nucleophilic at the β position and can be readily protonated there.
Addition of water to the intermediate cation gives another **hemiacetal**, which upon **elimination** gives the product aldehyde.

Recognize that this is not the final product, but is closely related to it

Deprotonation gives the aldehyde product.

Write down the structure of the final product

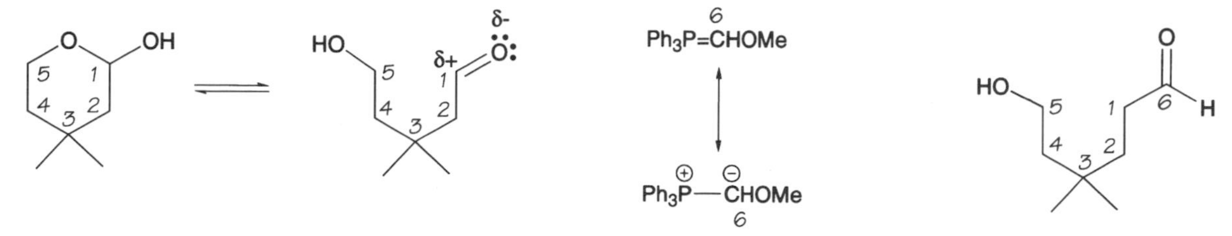

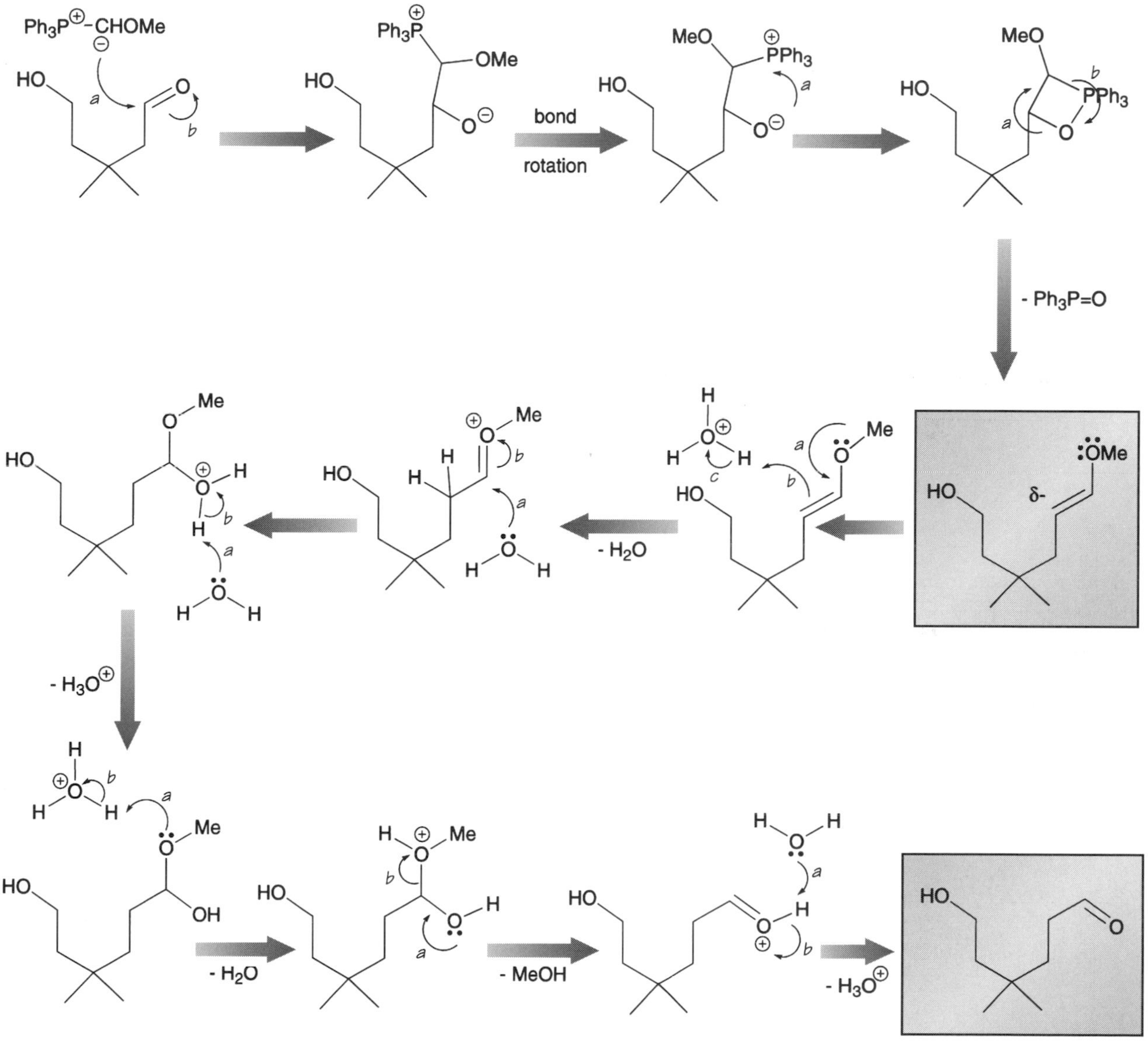

OH O 5 1 4 3 2

HO 5 4 3 2 1 δ+ δ- :Ö:

6
$Ph_3P{=}CHOMe$

$Ph_3\overset{\oplus}{P}{-}\overset{\ominus}{C}HOMe$
6

HO 5 4 3 2 1 6 O H

$Ph_3P^{\oplus}{-}\overset{\ominus}{C}HOMe$ HO a O b

$Ph_3\overset{\oplus}{P}$ OMe HO $O^{\ominus}$

bond rotation

MeO $\overset{\oplus}{P}Ph_3$ HO a $O^{\ominus}$

MeO HO b PPh_3 a O

- $Ph_3P{=}O$

HO δ- :ÖMe

H H $\overset{\oplus}{O}$ H c HO a b Ö Me

- H_2O

HO H H $\overset{\oplus}{O}$ Me b a H Ö H

HO O Me $\overset{\oplus}{O}$ H b H a H Ö H

- $H_3O^{\oplus}$

H $\overset{\oplus}{O}$ H H b a Ö Me HO OH

- H_2O

HO H $\overset{\oplus}{O}$ Me b Ö H a

- MeOH

H Ö H a HO O H $\oplus$ b

- $H_3O^{\oplus}$

HO O

Summary: Aldehydes or ketones and Wittig reagents react to give alkenes. The Wittig reagent $Ph_3P{=}CHOMe$ is especially useful as it gives another aldehyde product which is one carbon longer than the original compound.

Now try questions 3.18 and 3.19 at the end of this Chapter

3.7

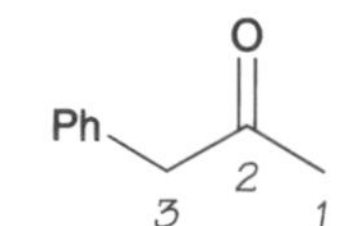

(i) NH_4Cl, KCN, H_2O
(ii) HCl, H_2O, heat
(iii) NaOH

H_2N CO_2H Ph

Label electrophilic/nucleophilic and acidic/basic sites of all reactants, and number identical atoms in the starting material and product

The carbonyl group is polarized by the *electronegativity* difference between carbon and oxygen, making the carbon electrophilic.
Cyanide is a good nucleophile and a weak base. Ammonium chloride generates ammonia in aqueous solution.

Identify the most reactive sites, if more than one exists

The ketone is the only electrophilic site.

Recall the characteristic reactions of the most reactive functional groups and, by considering the reaction conditions, decide which is the most appropriate

Ketones and amines react to give **imines**.

Work through the mechanism leading to the intermediate product

Nucleophilic **addition–elimination** of NH_3 to the ketone carbonyl group

Repeat the above four steps...

* Imines, like ketones, are polarized by the **electronegativity** difference between carbon and nitrogen.
Nucleophilic addition of cyanide occurs to give a tetrahedral intermediate.
* Acid-catalysed hydrolysis of the nitrile to an amide then occurs.
* NaOH is a good base and nucleophile; **hydrolysis** of the amide function occurs under basic conditions.

Recognize that this is not the final product, but is closely related to it

Not needed here.

Write down the structure of the final product

δ-
Ph 3 2 δ+ 1 (O)
K⊕ ⊖C≡N 4
H2N 4 CO2H Ph 3 2 1

$$NH_4^+ + H_2O \rightleftharpoons NH_3 + H_3O^+$$

Ph (ketone) + $H_3O^{\oplus}$ (a, b) → - H_2O → Ph, $H-O^{\oplus}$ (b), NH_3 (a) → Ph, HO, $NH_3^{\oplus}$ → - $H_3O^{\oplus}$ → Ph, HO, NH_2, H_2O (a, b)

↓

Ph, HO, NH_2, $H_3O^{\oplus}$ (a, b) → - H_2O → Ph, $H_2O^{\oplus}$, NH_2 (a, b) → - H_2O → Ph, $H-N^{\oplus}-H$ (b), ⊖CN (a) → [Ph, δ- N≡C δ+, NH_2]

↓

Ph, N≡C, NH_2, $H_3O^{\oplus}$ (a, b) → - H_2O → Ph, $H-N^{\oplus}$≡C, NH_2, H_2O (a, b) → Ph, H-N=C, $OH_2^{\oplus}$, NH_2, H_2O (a, b) → - $H_3O^{\oplus}$ → Ph, H—N=C(OH), NH_2

⇌

Ph, H_2N-C(=O δ-) δ+, NH_2 → Ph, HO⊖ (a), H_2N-C(=O), NH_2 (b) → + 2H⁺ → Ph, $H_3N^{\oplus}$, O-H, OH, NH_2, ⊖OH (a, b, c) → - NH_3, - H_2O → [Ph, HO-C(=O), NH_2]

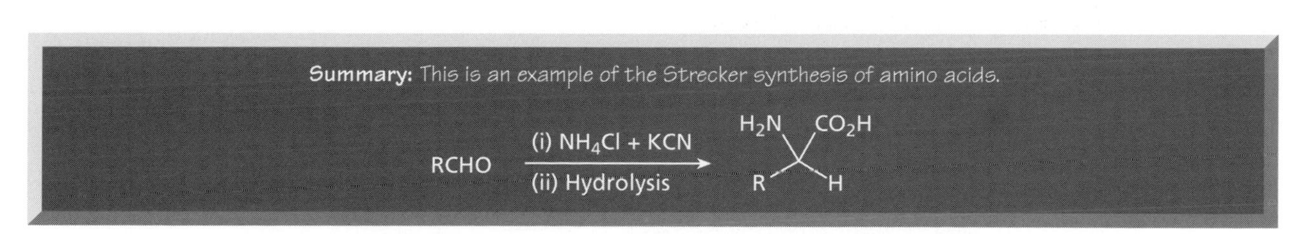
Summary: This is an example of the Strecker synthesis of amino acids.
RCHO
(i) NH4Cl + KCN
(ii) Hydrolysis
H2N CO2H
R H

Now try questions 3.20 and 3.21 at the end of this Chapter

Supplementary questions

3.8

(i) $NaHSO_3$, NaCN, H_2O

(ii) HCl, H_2O

3.9

H_2O, acetone, Δ

3.10

$LiAlH_4$, Et_2O

then acidic work-up

3.11

CF_3CO_2H, H_2O

+ PhCHO

3.12

(i) *t*-BuCHO, Et_3N

(ii) CH_3COCl, Et_3N

3.13

(i) ,THF, 25°C

(ii) HCl, H_2O

3.14

2-furaldehyde (CHO) —KCN→ furoin (O, OH)

3.15

PhCHO —(i) KCN, PhCOCl; (ii) NaOEt, piperonal (CHO)→ Ph–CO–CH(O_2CPh)–(3,4-methylenedioxyphenyl)

3.16

H–CO–CO–H —NaOH, then acidic work-up→ HO–CH_2–CO–OH

3.17

piperonal (CHO) —CH_2O, NaOH, then acidic work-up→ piperonyl alcohol (CH_2OH) + HCO_2H

3.18

Br–CH_2–CO_2Et —(i) $P(OEt)_3$; (ii) NaH; (iii) Ph–CO–Ph→ Ph_2C=CH–CO_2Et

3.19

Br–$(CH_2)_5$–$\overset{\oplus}{P}Ph_3$ Br^- —(i) 2 KO^tBu; (ii) PhCHO→ cyclopentylidene=CH–Ph

3.20

CH_3CHO —(i) NaCN, NH_4Cl; (ii) HCl, H_2O→ H_3C–CH(NH_2)–CO_2H

3.21

Phthalimide (NH) —(i) KH; (ii) Cl–CH(CH_2Ph)–CO_2Et; (iii) $NH_2NH_2.H_2O$→ H_2N–CH(CH_2Ph)–CO_2Et

4 Enolate chemistry

The α-protons of carbonyl groups are acidic due both to inductive withdrawal of the adjacent carbonyl group, which weakens the carbon–hydrogen bond, and to resonance stabilization of the enolate which is formed.

Enols and enolates are basic and nucleophilic, and participate in a variety of reactions:

Halogenation

- Can be acid- or base-catalysed—the Iodoform reaction.
- Hell–Vollhardt–Zelinsky reaction is used to prepare α-haloacid halides.

Alkylation

Some important aspects are:

- Nature of the base: this can be RO^- (R=Me, Et, *t*-Bu) or LDA ($[Me_2CH]_2NLi$).
- Nature of the electrophiles: the reaction is not suitable for 3° alkyl halides; for these substrates, Lewis acid-catalysed alkylation is used.
- The problem of O versus C alkylation (enolates are ambident nucleophiles).
- There are two very important syntheses which use the alkylation of enolates:
  (a) Malonic ester synthesis—alkylation followed by hydrolysis and decarboxylation gives substituted acetic acids:

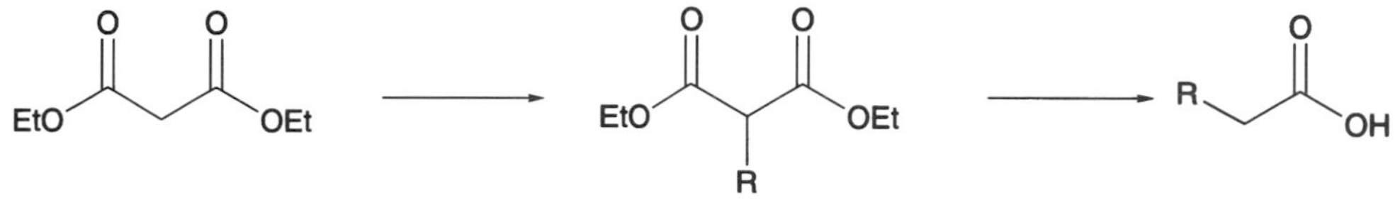

  (b) Acetoacetic ester synthesis—alkylation followed by hydrolysis and decarboxylation gives substituted acetones:

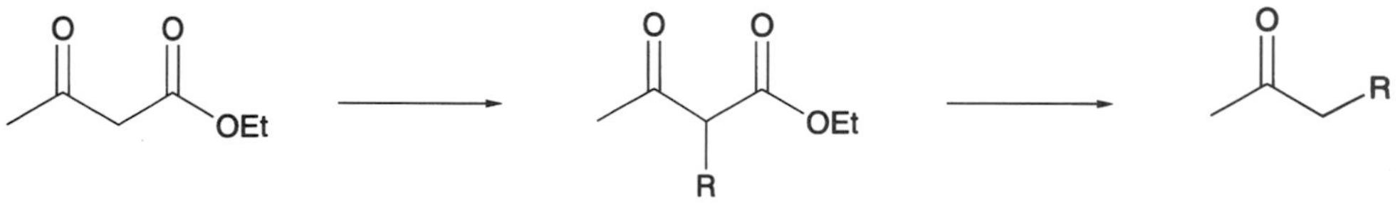

Condensation reactions

There are many examples of this process:

1 Aldol condensation

- Acid- and base-catalysed reactions.
- Mixed and intramolecular aldols.
- Dehydration to give α, β-unsaturated carbonyl derivatives.

2 Michael reaction and the Robinson ring annelation.

3 Claisen ester condensation, and other condensations. Perkin, Reformatsky, Stobbe, Mannich, Dieckmann, Darzens and Knoevenagel reactions.

4.1

4, 3, 2, 1, H, O

(i) KOH, H_2O

(ii) HCl, H_2O, Δ

O, H

Label electrophilic/nucleophilic and acidic/basic sites of all reactants, and number identical atoms in the starting material and product

The carbonyl group is polarized by the **electronegativity** difference between carbon and oxygen, making the carbon electrophilic. The α-protons of a carbonyl group are acidified by **inductive** withdrawal, and the resulting enolate is resonance stabilized.
KOH is a good base.

Identify the most reactive sites, if more than one exists

The aldehyde function is the only reactive group.

Recall the characteristic reactions of the most reactive functional groups and, by considering the reaction conditions, decide which is the most appropriate

Enolization of the aldehyde is followed by standard **nucleophilic addition** to another aldehyde carbonyl group.

Work through the mechanism leading to the intermediate product

Intermolecular **nucleophilic addition** gives condensation of two molecules of the aldehyde, to give a β-hydroxyaldehyde.

Repeat the above four steps...

The β-hydroxyaldehyde can be easily protonated on the hydroxyl oxygen by strong acid, and **elimination** to the α,β-unsaturated product occurs.

Recognize that this is not the final product, but is closely related to it

Not needed here.

Write down the structure of the final product

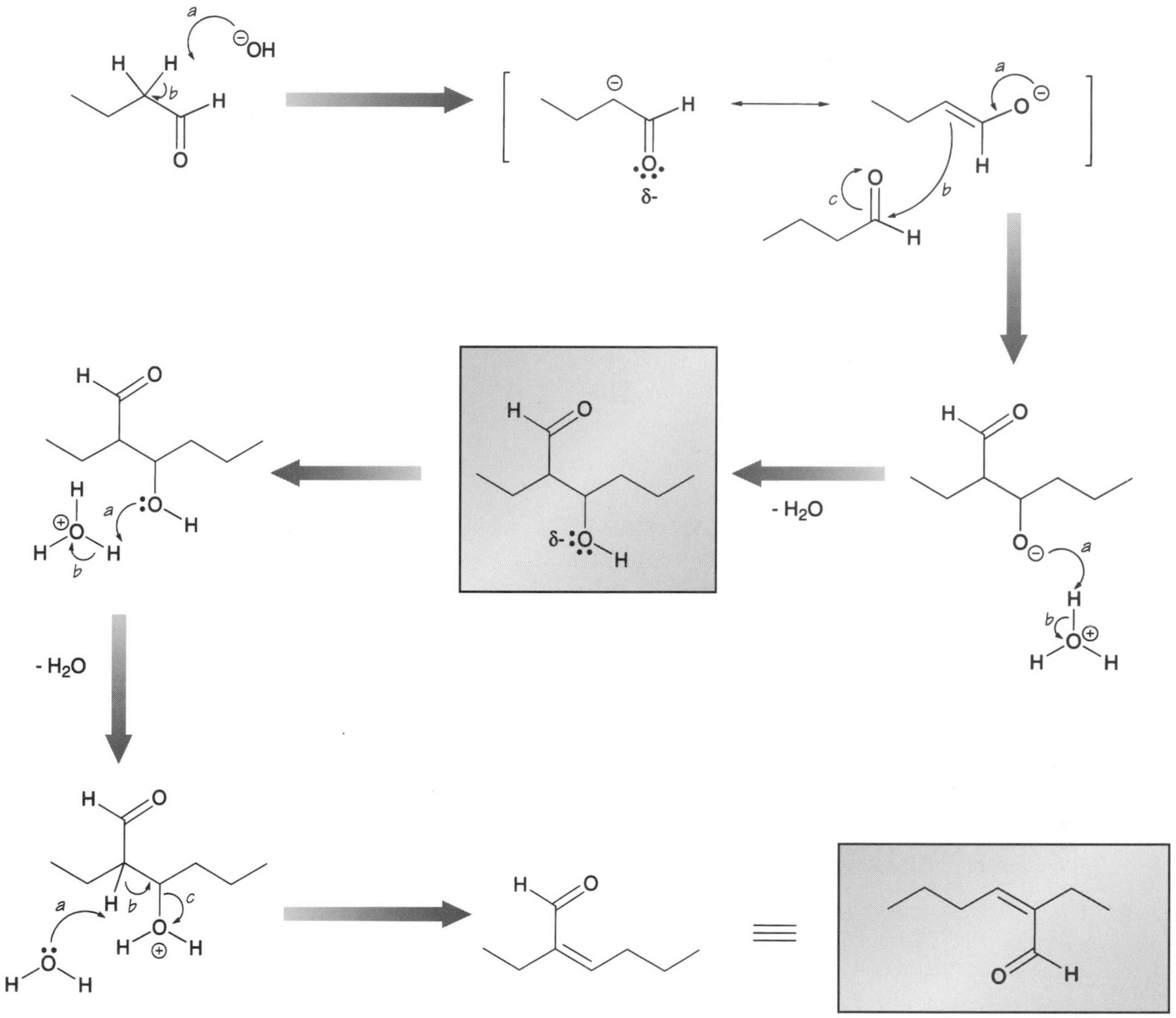

δ+ 1 H 4 2 3 O δ-

4 3 2 1 3 2 4 1 O H

$^{\ominus}OH$ a H H b H O

δ- O H ⊖ a O⊖ H b c O H

H O | H O | H O | - H_2O | δ- O H | O⊖ a H b O⊕ H H

H a O H ⊕O H H b

- H_2O

H O a H b c O H ⊕ H H O H | H O | O H

Summary: This is an example of an aldol reaction, which results in the condensation of two equivalents of an aldehyde to give a β-hydroxyaldehyde, which can be easily dehydrated.

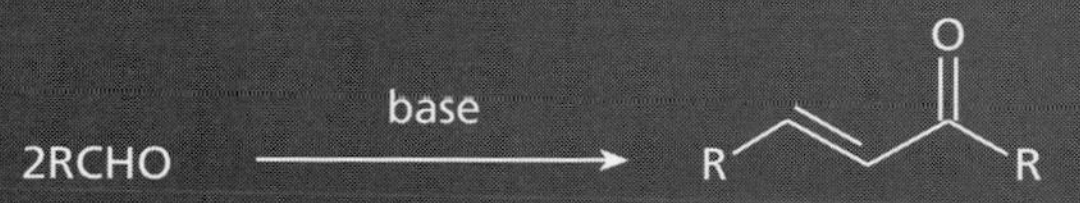

2RCHO —base→ R–CH=CH–C(=O)–R

Now try questions 4.8 and 4.9 at the end of this Chapter

4.2

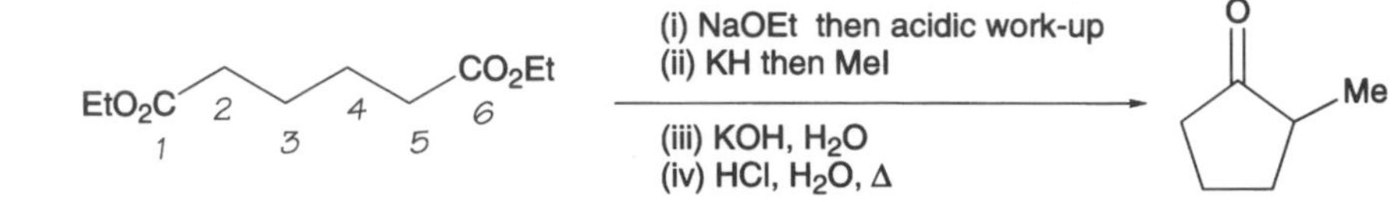

Label electrophilic/nucleophilic and acidic/basic sites of all reactants, and number identical atoms in the starting material and product

The carbonyl group is polarized by the **electronegativity** difference between carbon and oxygen, making the carbon electrophilic. The α-protons of a carbonyl group are acidified by **inductive** withdrawal, and the resulting enolate is resonance stabilized.
NaOEt is a good base.

Identify the most reactive sites, if more than one exists

C-2 and C-5 are the most acidic sites (pK_a approx. 25).

Recall the characteristic reactions of the most reactive functional groups and, by considering the reaction conditions, decide which is the most appropriate

Deprotonation of the α-protons of the ester is easily achieved by NaOEt to give an enolate, which is a good nucleophile. Esters readily undergo nucleophilic **addition–elimination**.

Work through the mechanism leading to the intermediate product

Intramolecular attack of the enolate at the other carbonyl gives a ketone.
One of the products (ethoxide) is a good base, capable of removing a proton in the product and driving the **equilibrium** to the right.

Repeat the above four steps...

* MeI is a good electrophile, by virtue of the electronegativity difference of carbon and iodine; **nucleophilic substitution** by the ketone enolate gives the methylated product.
* KOH is a good nucleophile and base; ester hydrolysis by **addition-elimination** gives the corresponding carboxylate anion.

Recognize that this is not the final product, but is closely related to it

Acidification and heating causes **decarboxylation**, to give the final product.

Write down the structure of the final product

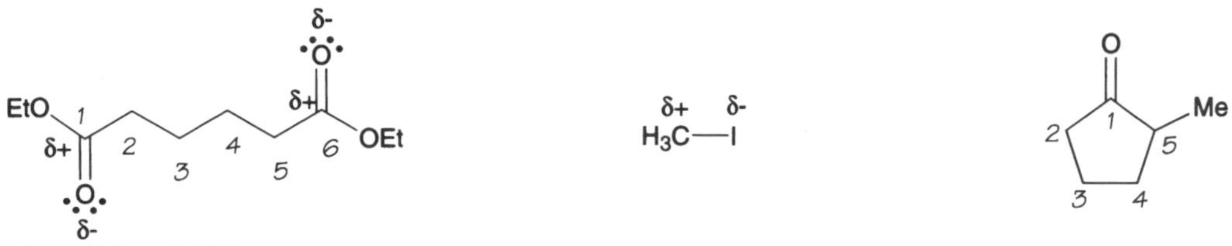

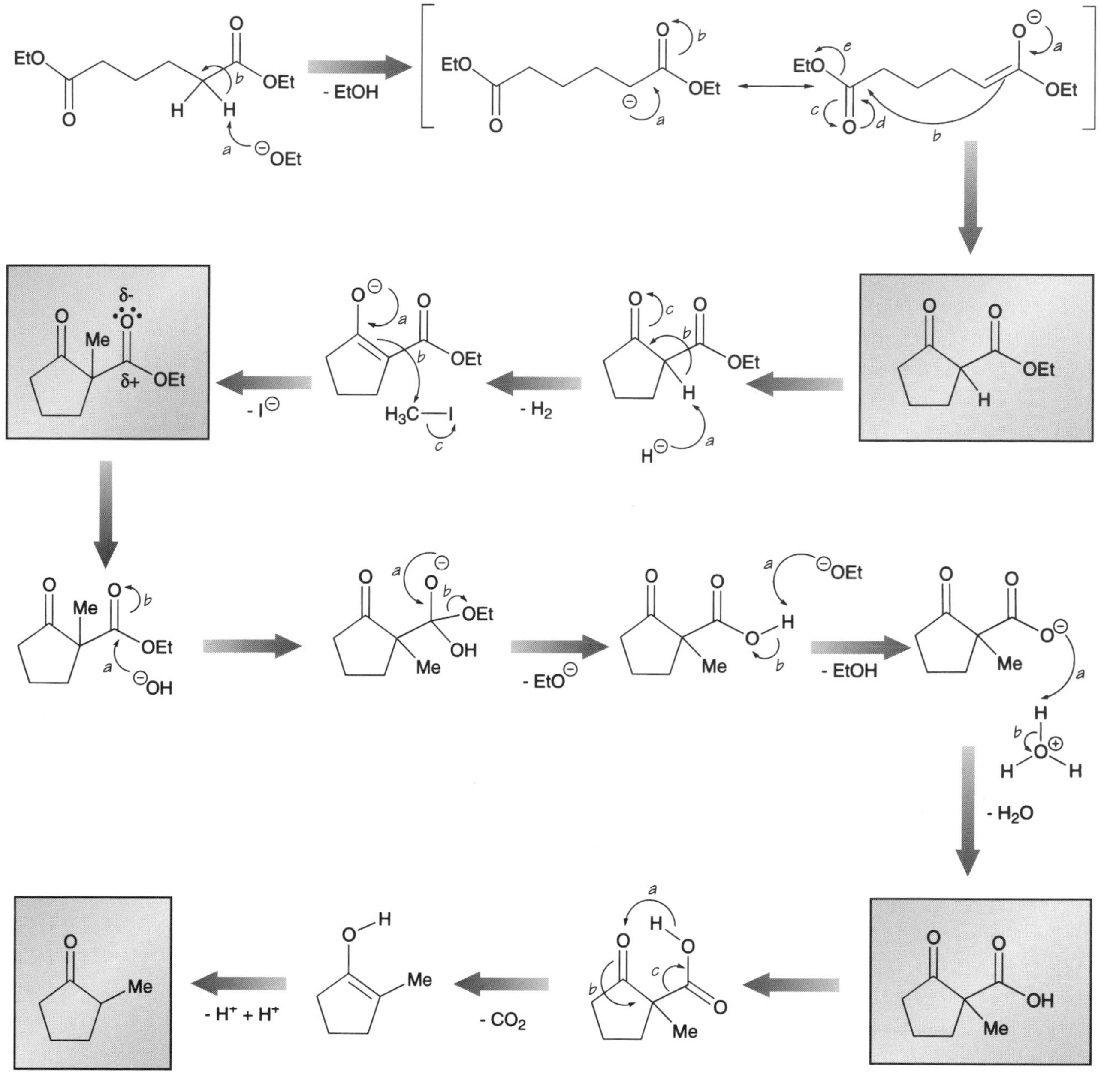

Summary: This is an example of the Dieckmann cyclisation. Hydrolysis and decarboxylation of the products gives a cyclic ketone.

$EtO_2C\text{–}(CH_2)_n\text{–}CO_2Et \xrightarrow{NaOEt}$ cyclic ketone $(CH_2)_n$

Now try questions 4.10 and 4.11 at the end of this Chapter

4.3

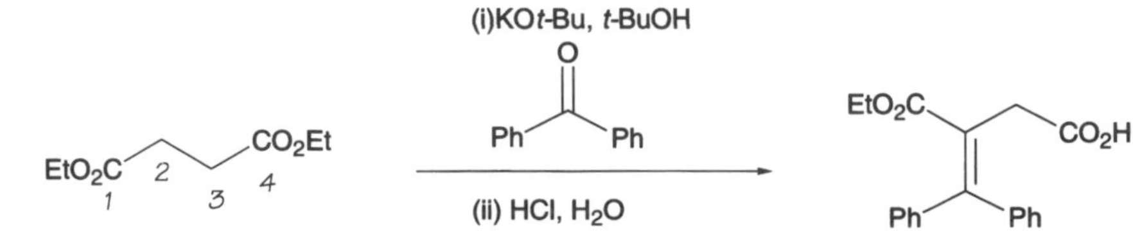

Label electrophilic/nucleophilic and acidic/basic sites of all reactants, and number identical atoms in the starting material and product

The carbonyl group is polarized by the **electronegativity** difference between carbon and oxygen, making the carbon electrophilic. The α-protons of a carbonyl group are acidified by **inductive** withdrawal, and the resulting enolate is resonance stabilized.
KOt-Bu is a very strong base, but a poor nucleophile, due to steric hindrance.

Identify the most reactive sites, if more than one exists

C-2 and C-3 are the most acidic sites.

Recall the characteristic reactions of the most reactive functional groups and, by considering the reaction conditions, decide which is the most appropriate

Ester enolates are good nucleophiles; ketones are good electrophiles, and readily undergo **nucleophilic addition**.

Work through the mechanism leading to the intermediate product

Nucleophilic addition of the ester enolate to the ketone.

Repeat the above four steps...

The alkoxide so generated will undergo a **nucleophilic-addition** reaction with the terminal ester. A five-membered **lactone** is formed on thermodynamic grounds.
Strong acid protonates the lactone carbonyl oxygen, converting it to a good **leaving group**; elimination then occurs to give a highly stabilized **carbocation** intermediate.

Recognize that this is not the final product, but is closely related to it

Acidification on work-up gives the carboxylic acid product.

Write down the structure of the final product

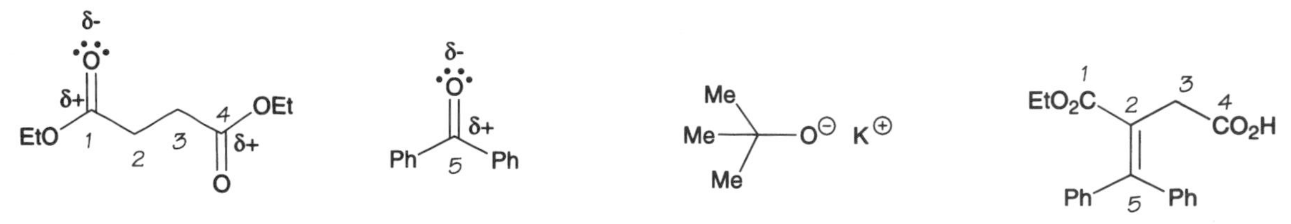

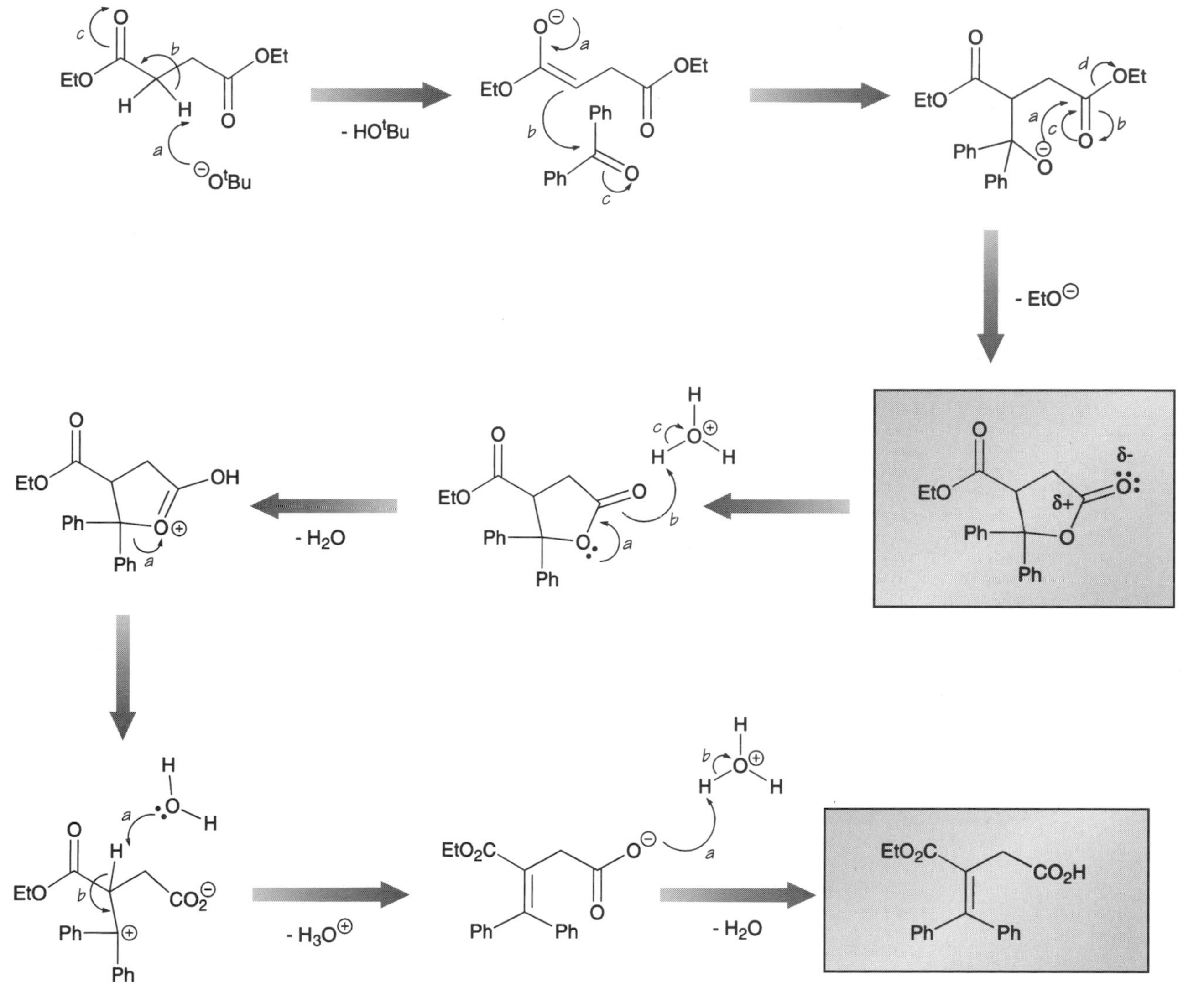

Summary: This is an example of the Stobbe reaction; noteworthy in this reaction is that the condensation proceeds with concomitant hydrolysis of the terminal ester.

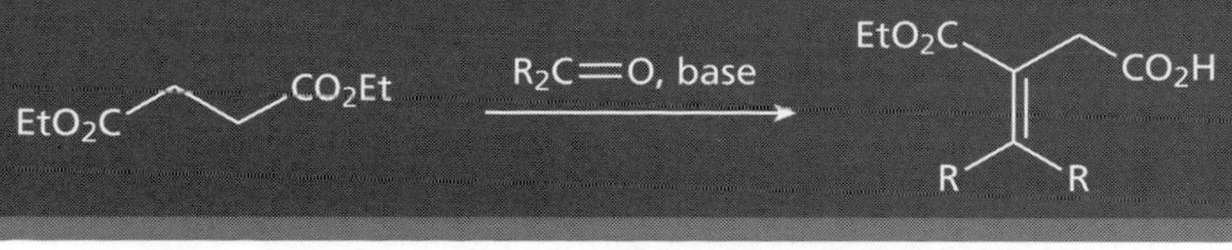

$$EtO_2C\text{-}CH_2CH_2\text{-}CO_2Et \xrightarrow{R_2C{=}O,\ base} EtO_2C\text{-}C(=CR_2)\text{-}CH_2\text{-}CO_2H$$

Now try questions 4.12 and 4.13 at the end of this Chapter

4.4

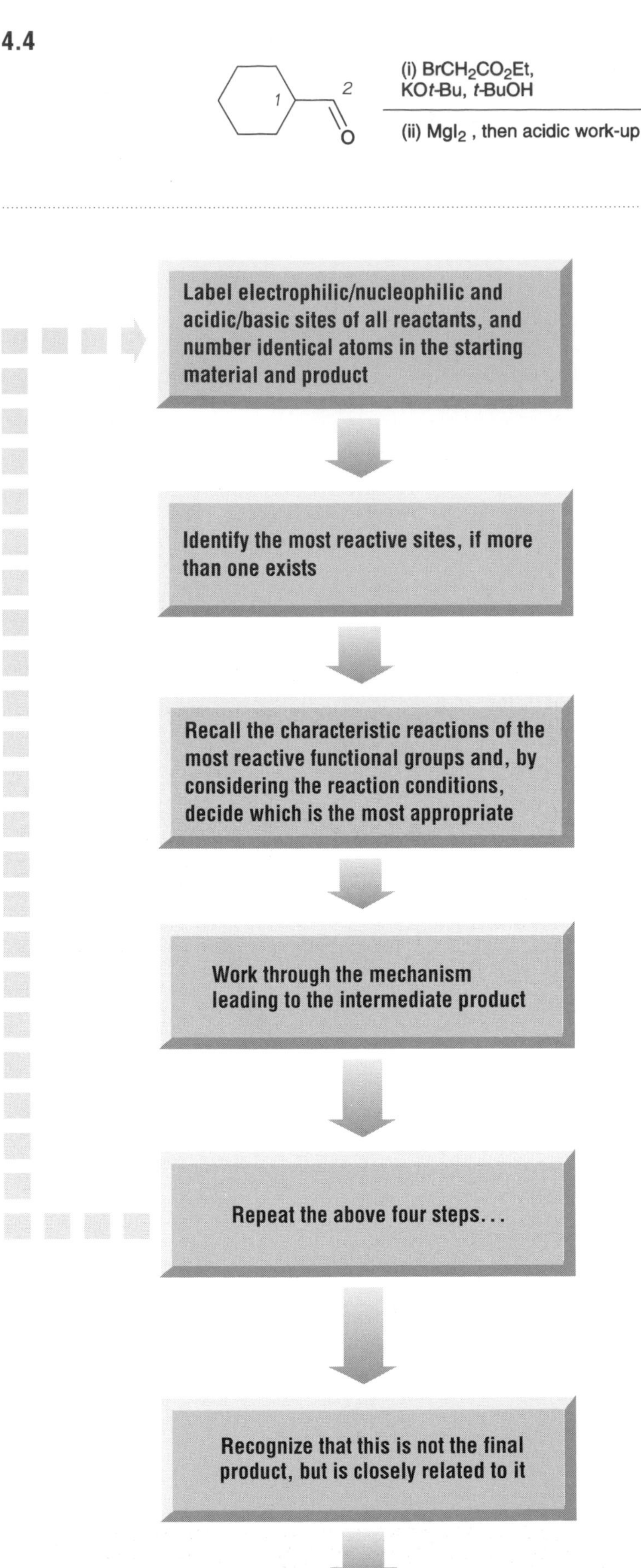

(i) $BrCH_2CO_2Et$, KO*t*-Bu, *t*-BuOH
(ii) MgI_2, then acidic work-up
1
2
O
I
CO_2Et
OH
Label electrophilic/nucleophilic and acidic/basic sites of all reactants, and number identical atoms in the starting material and product
Identify the most reactive sites, if more than one exists
Recall the characteristic reactions of the most reactive functional groups and, by considering the reaction conditions, decide which is the most appropriate
Work through the mechanism leading to the intermediate product
Repeat the above four steps...
Recognize that this is not the final product, but is closely related to it
Write down the structure of the final product

The carbonyl group is polarized by the **electronegativity** difference between carbon and oxygen, making the carbon electrophilic. The α-protons of a carbonyl group are acidified by **inductive** withdrawal, and the resulting enolate is resonance stabilized.
KOt-Bu is a very strong base, but a poor nucleophile, due to steric hindrance.

The aldehyde is the only electrophilic site.

Aldehydes are susceptible to **nucleophilic addition**.
The α-hydrogens of a bromoester are especially acidic, and can easily be removed by **t**-butoxide, giving an enolate which is nucleophilic.

Nucleophilic addition to the aldehyde gives an alkoxide, which readily undergoes an intramolecular **nucleophilic substitution** at the adjacent bromide to give an **epoxide** product.

MgI_2 is a good Lewis acid, which will coordinate to the electron-rich epoxide oxygen, and catalyse **nucleophilic substitution** of the epoxide by iodide; the regioselectivity is given by the stability of the carbocation which is generated if an S_N1-like mechanism is assumed.

Acidic work-up protonates the alkoxide, to give the product.

Summary: This is an example of the Darzens reaction.

$$XCH_2CO_2Et + R_2C{=}O \xrightarrow{\text{base}} \text{epoxide: } R_2C(O)CH\text{–}CO_2Et$$

X=Br, Cl

Now try questions 4.14 and 4.15 at the end of this Chapter

4.5

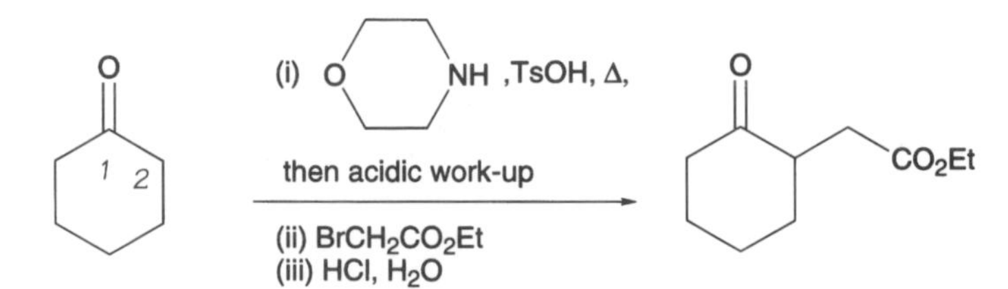

Label electrophilic/nucleophilic and acidic/basic sites of all reactants, and number identical atoms in the starting material and product

The carbonyl group is polarized by the **electronegativity** difference between carbon and oxygen, making the carbon electrophilic. The α-protons of a carbonyl group are acidified by **inductive** withdrawal, and the resulting enolate is resonance stabilized.
Morpholine is a good nucleophile, but not basic enough to deprotonate a ketone.

Identify the most reactive sites, if more than one exists

The ketone is the only electrophilic site.

Recall the characteristic reactions of the most reactive functional groups and, by considering the reaction conditions, decide which is the most appropriate

Ketones and amines react to give imines.

Work through the mechanism leading to the intermediate product

The ketone is protonated, and **nucleophilic addition** of morpholine to the ketone then proceeds, but in this case only **elimination** of the intermediate to an **enamine** is possible.

Repeat the above four steps...

An enamine is an excellent nucleophile (compare with an enol) at the β-carbon. Ethyl bromoacetate is a good electrophile, with Br^- leaving group, and **nucleophilic substitution** then occurs.

Recognize that this is not the final product, but is closely related to it

Acid-catalysed iminium ion hydrolysis (compare hydrolysis of an acetal); ester hydrolysis does not occur under these mild conditions.

Write down the structure of the final product

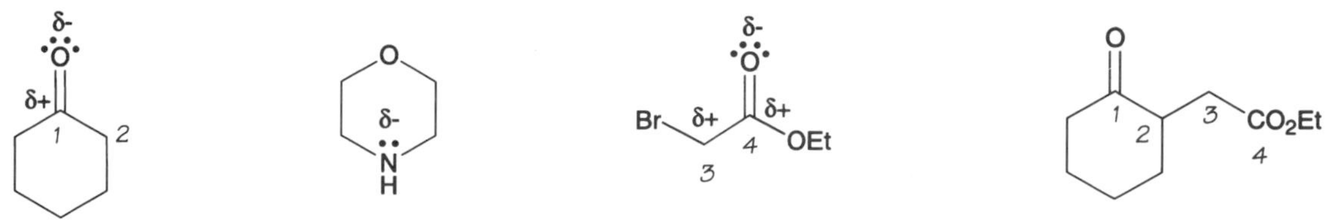

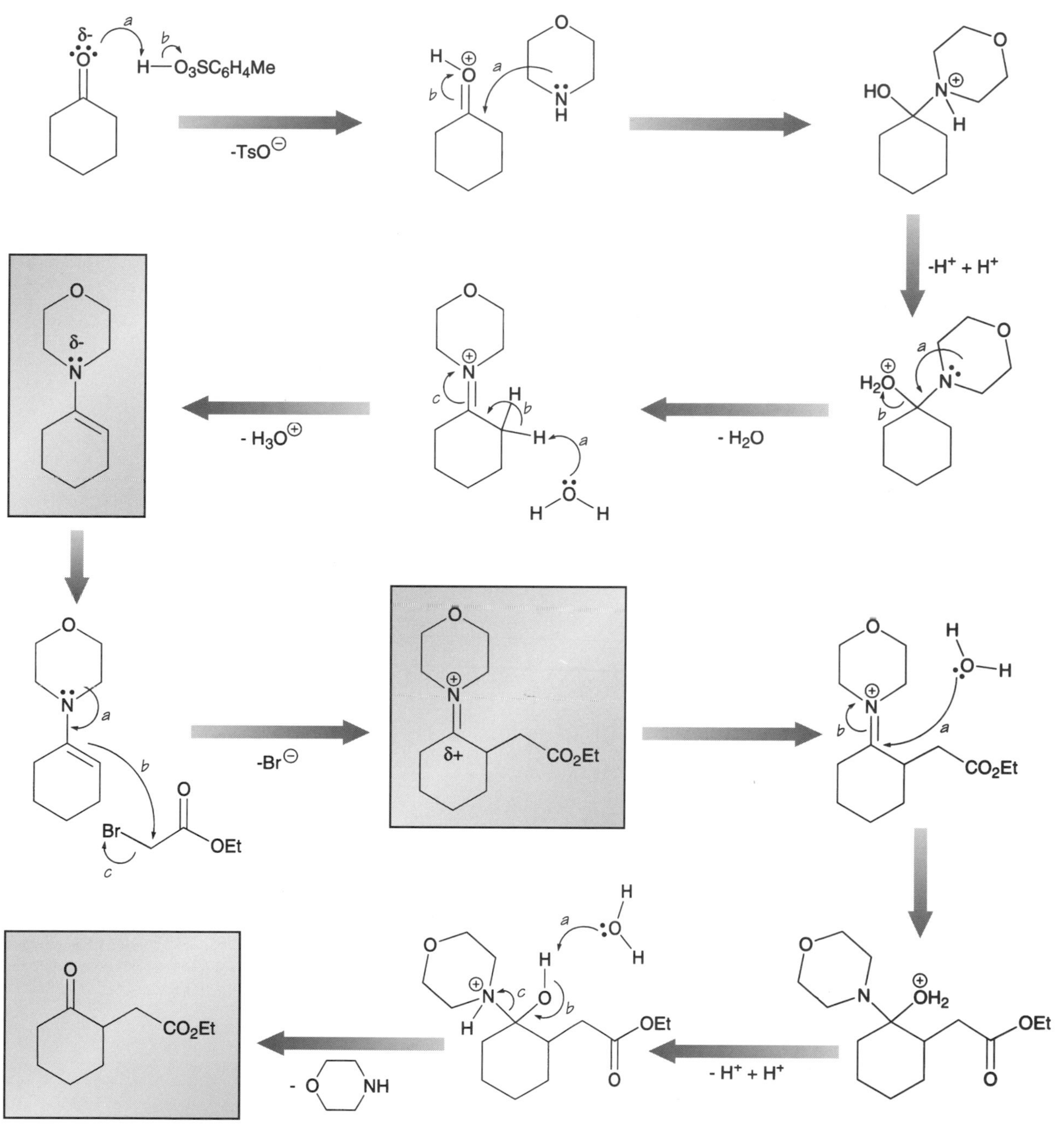

-$TsO^{\ominus}$

$-H^+ + H^+$

$- H_2O$

$- H_3O^{\oplus}$

$-Br^{\ominus}$

$- H^+ + H^+$

Summary: This is an example of the Stork enamine synthesis; this reaction is equivalent to the alkylation of an enolate.

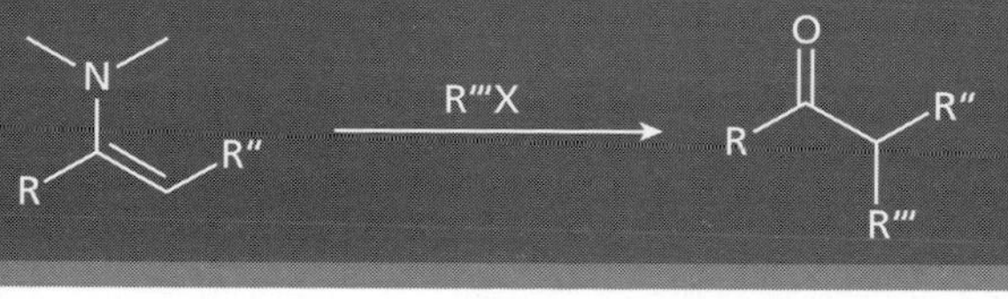

R‴X

Now try questions 4.16 and 4.17 at the end of this Chapter

4.6

4 CHO, 5, 6, CHO 7 + HO_2C–(1)–C(=O)(2)–(3)–CO_2H $\xrightarrow{MeNH_2,\ H_2O}$ bicyclic ketone (N–Me bridge)

Label electrophilic/nucleophilic and acidic/basic sites of all reactants, and number identical atoms in the starting material and product

Aldehydes are good electrophiles, since the carbonyl group is polarized by the electronegativity difference between carbon and oxygen.
Methylamine is a good nucleophile.
1,3-Dicarbonyl compounds are acidic at the α position (here labelled C-1 and C-3).

Identify the most reactive sites, if more than one exists

The dialdehyde is very reactive, and susceptible to **nucleophilic addition** at C-4 and C-7.
Acetone dicarboxylic acid is a good nucleophile at C-3, and exists in equilibrium with its enolic **tautomer**.
Dimethylamine is a good nucleophile and base.

Recall the characteristic reactions of the most reactive functional groups and, by considering the reaction conditions, decide which is the most appropriate

Aldehydes react with amines to form **imines**.

Work through the mechanism leading to the intermediate product

Addition–elimination reaction of the amine generates an **iminium** cation, which is resonance stabilized.

Repeat the above four steps...

The **iminium** cation is easily attacked by the nucleophilic enol; this whole process is repeated using the other aldehyde group and the alternative enol, to form a bicyclic product.

Recognize that this is not the final product, but is closely related to it

Double **decarboxylation** of the carboxylate anion gives the product.

Write down the structure of the final product

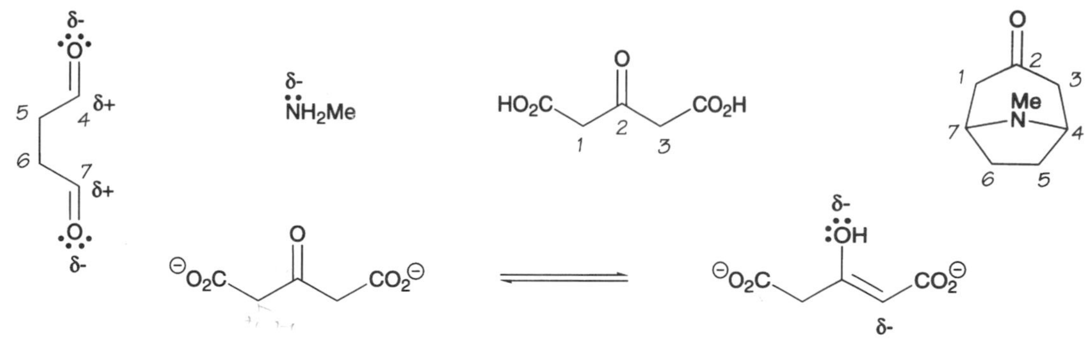

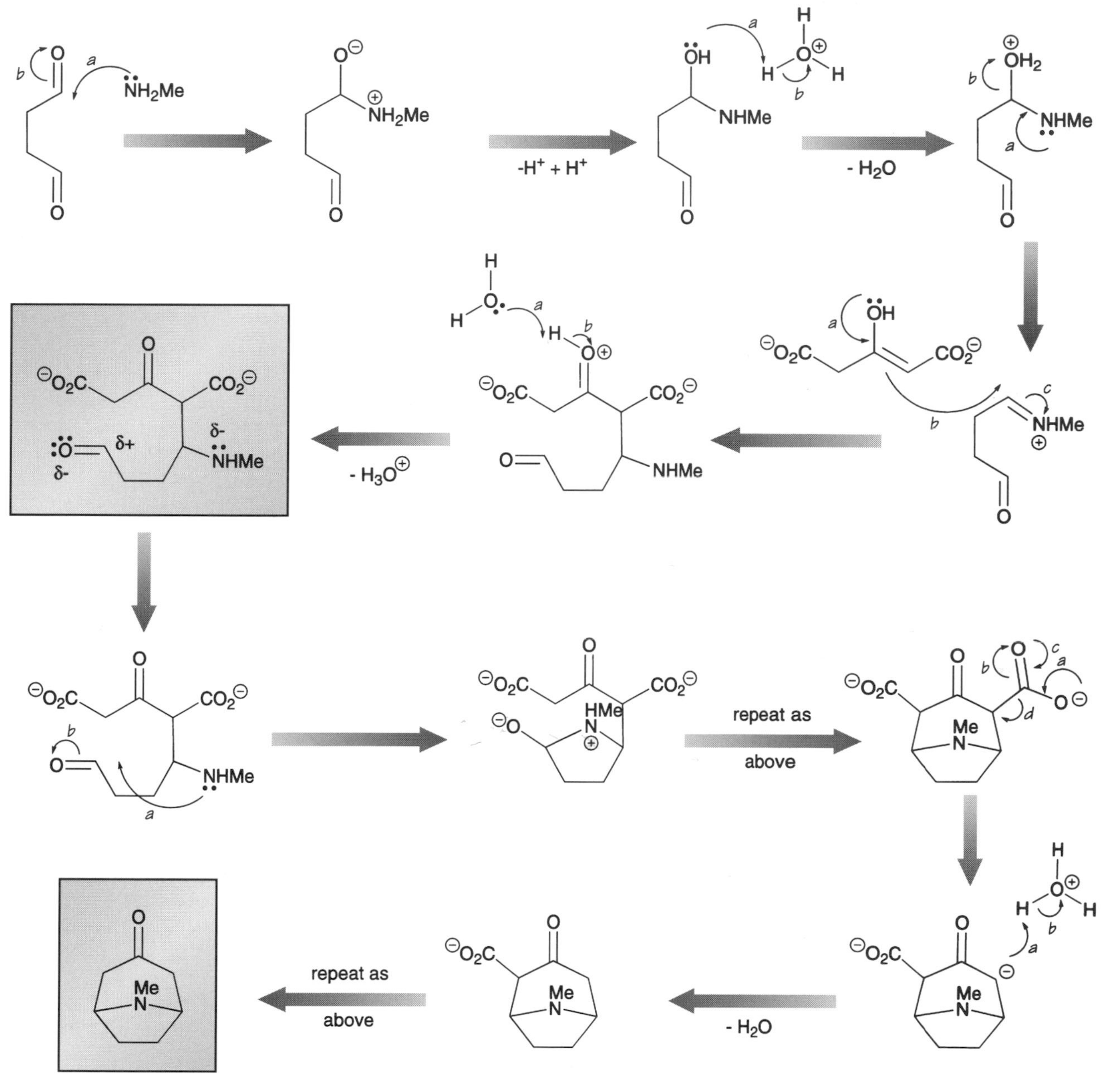

Summary: This is an example of the Mannich reaction.

Now try questions 4.18 and 4.19 at the end of this Chapter

4.7

EtO–CO–CH_2(6)–CO–OEt + Me(1)–C(2)(=O)–CH(3)=C(4)(Me)–Me(5) → (i) NaOEt; (ii) HCl, H_2O, Δ → 5,5-dimethylcyclohexane-1,3-dione

Label electrophilic/nucleophilic and acidic/basic sites of all reactants, and number identical atoms in the starting material and product

The α-protons of β-dicarbonyl compounds are highly acidic (pK_a about 13).
α,β-Unsaturated ketones are **polarized** at C-2 and C-4 by conjugation.
EtO− is a good base.

Identify the most reactive sites, if more than one exists

α,β-Unsaturated ketones are electrophilic at the β position (here labelled C-4). The α position of the β-dicarbonyl enolate is the most nucleophilic (here labelled C-6).

Recall the characteristic reactions of the most reactive functional groups and, by considering the reaction conditions, decide which is the most appropriate

α,β-Unsaturated carbonyl compounds readily undergo 1,4-nucleophilic addition reactions with enolate nucleophiles.

Work through the mechanism leading to the intermediate product

Repeat the above four steps...

Conjugate addition of the enolate to the *Michael acceptor*, followed by **equilibration** of the ketone enolate to the alternative (less stable) enolate; ketone formation then follows by standard **addition–elimination** mechanism at an ester. The overall reaction is under thermodynamic control, and proceeds because of the irreversible ring-closure step.

Acid-catalysed **hydrolysis** of the ester, followed by standard **decarboxylation** of the resulting β-ketoacid.

Recognize that this is not the final product, but is closely related to it

Recognize that the intermediate product can be converted to the final product by *tautomerization*. This can be catalysed by any available acid or base.

Write down the structure of the final product

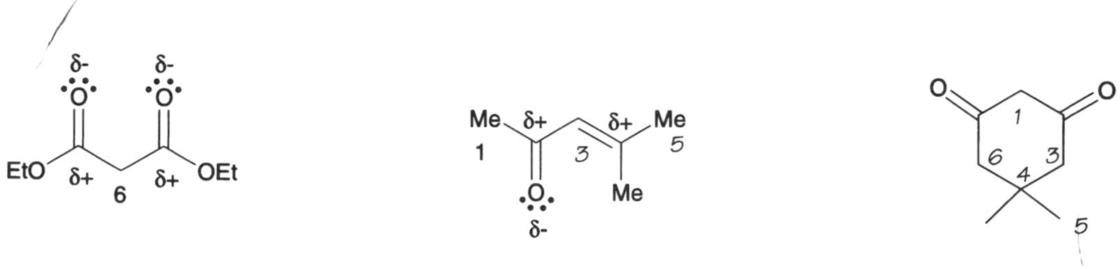

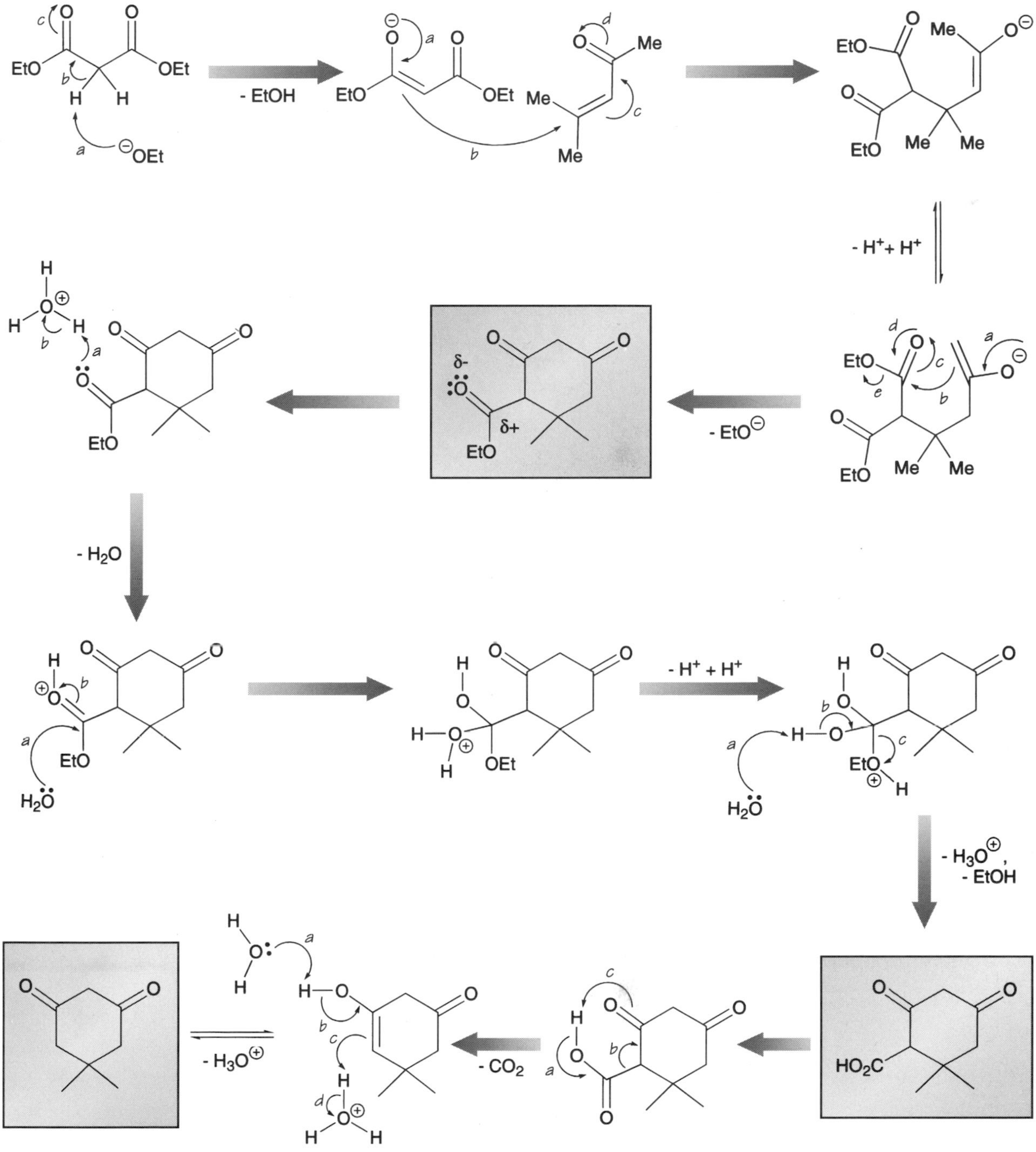

Summary: This is an example of a conjugate addition of an enolate (Michael reaction), followed by an intramolecular Claisen-type acylation.

Now try questions 4.20 and 4.21 at the end of this Chapter

Supplementary questions

4.8

EtO_2C CO_2Et

Piperidine; HOAc, EtOH, Δ

CO_2Et

4.9

NaOH, H_2O

4.10

KOH, py

then acidic work-up

4.11

(i) NaOMe then MeI
(ii) KOH/EtOH, Δ
(iii) HCl, H_2O

4.12

HO_2C CO_2H

(i) PhCHO, piperidine

(ii) HCl, H_2O

4.13

(i) NaH, $(EtO)_2CO$
(ii) NaH, $PhCH_2Br$

4.14

O O OEt

(i) PCl_5

(ii) $LiAlH_4$, then acidic work-up
(iii) PCl_3

Cl Cl

4.15

CO_2H O

HCl, H_2O

O

4.16

O CO_2Et

(i) N H, TsOH, Δ

(ii) CH_2=$CHCO_2Me$
(iii) HCl, H_2O

O CO_2Me CO_2Et

4.17

O N O Ph

(i) N H

(ii) H_2C=$CHCOCH_3$
(iii) HOAc, H_2O

O N O Ph

4.18

NMe_2 N H

(i) $(EtO_2C)_2CHNHAc$ / NaOEt

(ii) NaOH, H_2O, Δ
(iii) HCl, H_2O, Δ

H_2N CO_2H N H

4.19

O NMe_2

(i) NaOEt, O O OEt

(ii) NaOH, Δ
(iii) HCl, Δ

O

4.20

(i) NaOEt (cat.), EtOH

(ii) [but-1-en-3-one, Me]

(iii) HCl, H_2O

CO_2Et ... CO_2Et, O, Me

4.21

(i) $CH_2(CO_2Et)_2$, NaOEt, EtOH

(ii) KOH, H_2O, Δ

(iii) HCl, H_2O, Δ

O ... O, CO_2Et

5 Aromatic chemistry

Bonding in aromatic compounds

- The molecular orbital description of bonding in benzene derivatives and the importance of the delocalization of electrons. The concept of resonance stabilization, and the $(4n + 2)$ rule.
- The stability and reactivity of benzene derivatives is dominated by their aromaticity and the fact that the electronic effect of substituents can be transmitted around the ring by resonance:

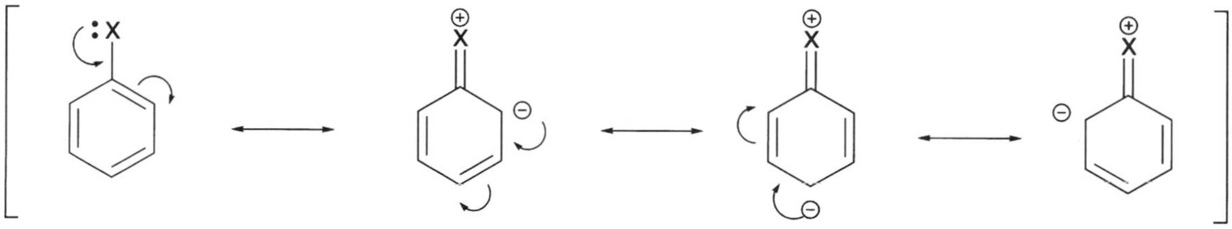

Electrophilic aromatic substitution (S_EAr)

1 Generalized mechanism: σ-complexes, π-complexes, the role of Lewis acids and aromatization as a driving force for substitution reactions.

2 Substitution of monofunctionalized benzene derivatives

- Orientation and reactivity is governed by the nature of ring substituents (inductive and mesomeric effects).
- The importance of kinetic and thermodynamic control.

3 Specific reaction and reagent examples:

- Deuteriation: D_2SO_4/D_2O
- Nitration: HNO_3/H_2SO_4
- Halogenation: $Br_2/FeBr_3$
- Sulfonation: SO_3/H_2SO_4
- Diazo coupling: PhN_2^+
- Friedel–Crafts alkylation and acylation: RCl, $AlCl_3$ and RCOCl, $AlCl_3$
- Vilsmeier formylation: $POCl_3$, Me_2NCHO

4 Special case: the Reimer–Tiemann reaction.

Nucleophilic aromatic substitution (S_NAr)

There are three possible mechanisms depending on the reaction conditions and the nature of the substrate:

1 S_NAr—needs activation with electron-withdrawing groups.

2 S_N1—diazonium salts react with nucleophiles.

3 Benzyne mechanism—aryne intermediates formed by the elimination of aryl halides.

Oxidation

Alkyl benzenes are easily oxidized to benzoic acid.

5.1

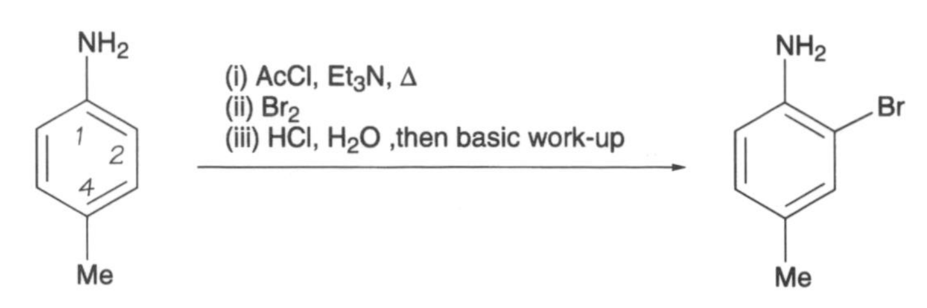

Label electrophilic/nucleophilic and acidic/basic sites of all reactants, and number identical atoms in the starting material and product

Amines are good nucleophiles (nitrogen possesses a lone pair) and carboxylic acid chlorides are reactive electrophiles by virtue of the electronegativity of chlorine and the fact that Cl^- is a good leaving group.

Identify the most reactive sites, if more than one exists

The aniline nitrogen is the most nucleophilic site, although the lone pair is linked by resonance to the aromatic ring, placing a $\delta-$ charge on the *o*- and *p*- positions (labelled C-2 and C-4 here).

Recall the characteristic reactions of the most reactive functional groups and, by considering the reaction conditions, decide which is the most appropriate

Reaction of an acid chloride with an amine nucleophile gives an amide product.

Work through the mechanism leading to the intermediate product

Addition–elimination of acetyl chloride with the aniline and expulsion of chloride.

Repeat the above four steps...

* Bromine is a good electrophile, and acetanilide a good nucleophile, due to the presence of the donating amide nitrogen. **Electrophilic aromatic substitution** by the bromine gives the aryl bromide product.
* Acid-catalysed amide **hydrolysis** regenerates the amine, but under the conditions of the hydrolysis reaction this is protonated, to give the anilinium salt as the product.

Recognize that this is not the final product, but is closely related to it

A basic work-up allows isolation of the free amine.

Write down the structure of the final product

$- Cl^{\ominus}$

$- Et_3NH^+$

$- Br^{\ominus}$

$- H_2O$

$-H^+ + H^+$

$- CH_3CO_2H$

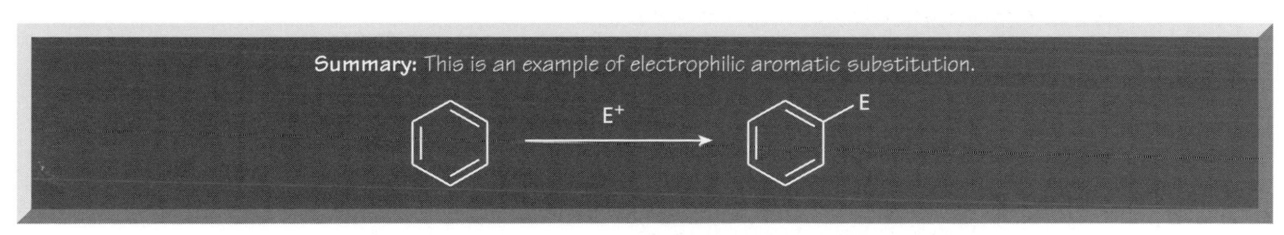
Summary: This is an example of electrophilic aromatic substitution.
E+
E

Now try questions 5.8 and 5.9 at the end of this Chapter

5.2

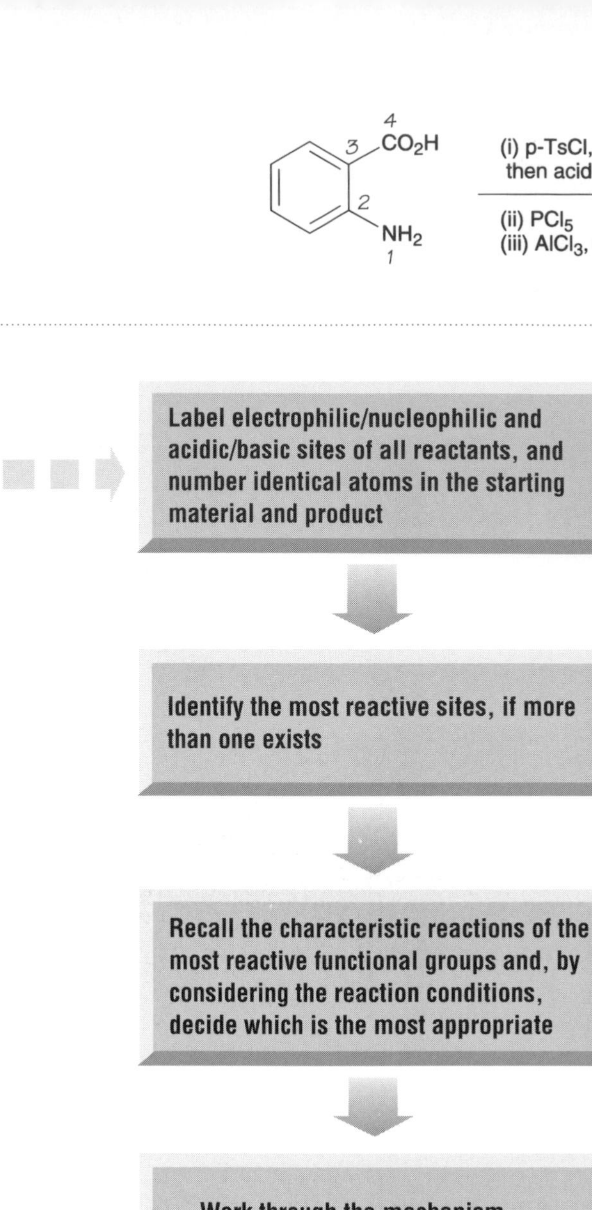

(i) p-TsCl, Na_2CO_3, then acidic work-up
(ii) PCl_5
(iii) $AlCl_3$, C_6H_6, Δ

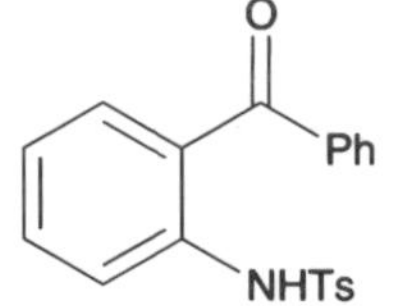

Label electrophilic/nucleophilic and acidic/basic sites of all reactants, and number identical atoms in the starting material and product

Carboxylic acids and amines are good nucleophiles (both oxygen and nitrogen possess lone pairs).
p-TsCl is an (electrophilic) acid chloride derived from a sulfonic acid.

Identify the most reactive sites, if more than one exists

The amine is the most nucleophilic site, since nitrogen is less electronegative than oxygen, even though in this case the carboxylic acid is deprotonated under the basic conditions.

Recall the characteristic reactions of the most reactive functional groups and, by considering the reaction conditions, decide which is the most appropriate

Reaction of an acid chloride with an amine nucleophile gives an amide product.

Work through the mechanism leading to the intermediate product

The aromatic amine undergoes a **nucleophilic addition–elimination** reaction at the sulfonic acid group, with loss of chloride; this generates a sulfonamide. The basic conditions of this reaction give a carboxylate product, which is reprotonated on acidic work-up.

Repeat the above four steps...

* Phosphorus pentachloride is an electrophile, with chloride **leaving groups**. The carboxylic acid reacts with PCl_5 to give an acid chloride, *via* an activated phosphoryl intermediate.
* The acid chloride reacts with the powerful Lewis acid, $AlCl_3$, to give an *acylium* cation (resonance-stabilized), which is very electrophilic and reacts with the good nucleophile, benzene, in an **electrophilic substitution** reaction.

Recognize that this is not the final product, but is closely related to it

Loss of a proton regenerates the aromatic system.

Write down the structure of the final product

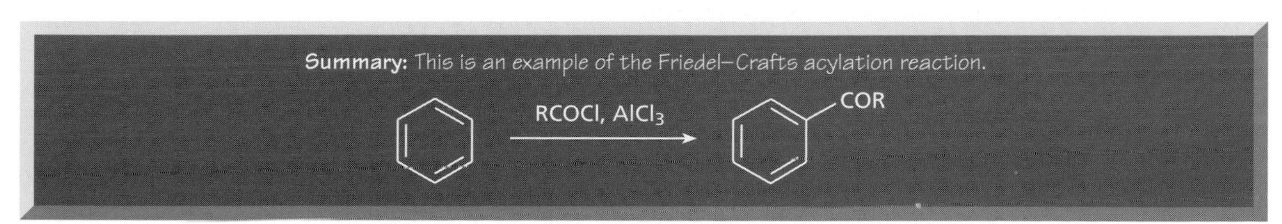

Now try questions 5.10 and 5.11 at the end of this Chapter

5.3

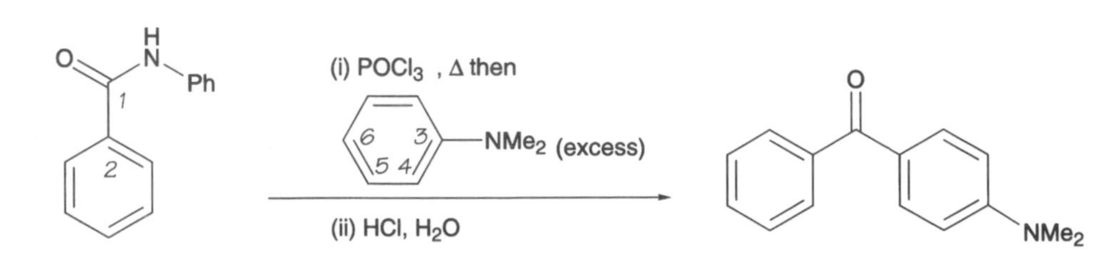

Label electrophilic/nucleophilic and acidic/basic sites of all reactants, and number identical atoms in the starting material and product

Amines are good nucleophiles, since nitrogen possesses a lone pair.
Amides are nucleophilic at oxygen and nitrogen.
$POCl_3$ is a potent electrophile, since there are three electron-withdrawing chlorines and each of these is also a good leaving group.

Identify the most reactive sites, if more than one exists

Amides are most nucleophilic on oxygen (the product has more resonance structures than that from reaction at nitrogen). Anilines are most nucleophilic at the *o*- and *p*- positions, but the latter is more sterically accessible.

Recall the characteristic reactions of the most reactive functional groups and, by considering the reaction conditions, decide which is the most appropriate

Amides react readily with $POCl_3$ to generate a chloromethyliminium salt; this is a potent electrophile, and reacts with aromatic compounds by an S_EAr reaction.

Work through the mechanism leading to the intermediate product

Initial *O*-phosphorylation of the amide by $POCl_3$ (to form a strong P–O bond) is followed by **addition–elimination** of Cl- to give a chloromethyliminium ion.

Repeat the above four steps...

* Electrophilic aromatic substitution at the *p*- position of the aniline with the chloromethyliminium electrophile gives an imine product.
* Acid-catalysed imine **hydrolysis** gives the ketone product, by an **addition–elimination** mechanism.

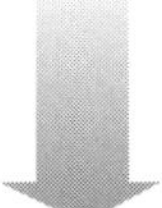

Recognize that this is not the final product, but is closely related to it

Final deprotonation of the **oxonium** cation gives the ketone product.

Write down the structure of the final product

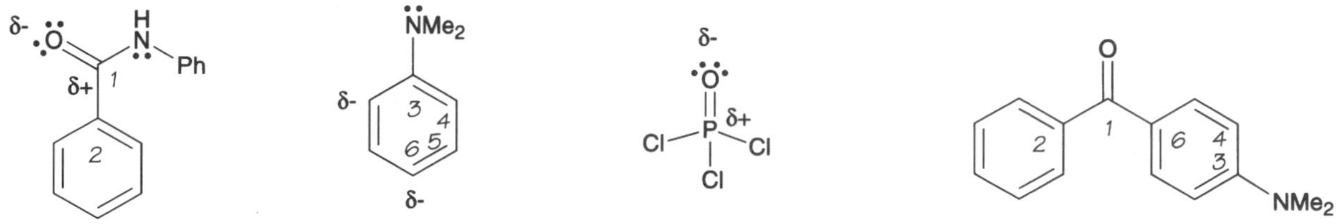

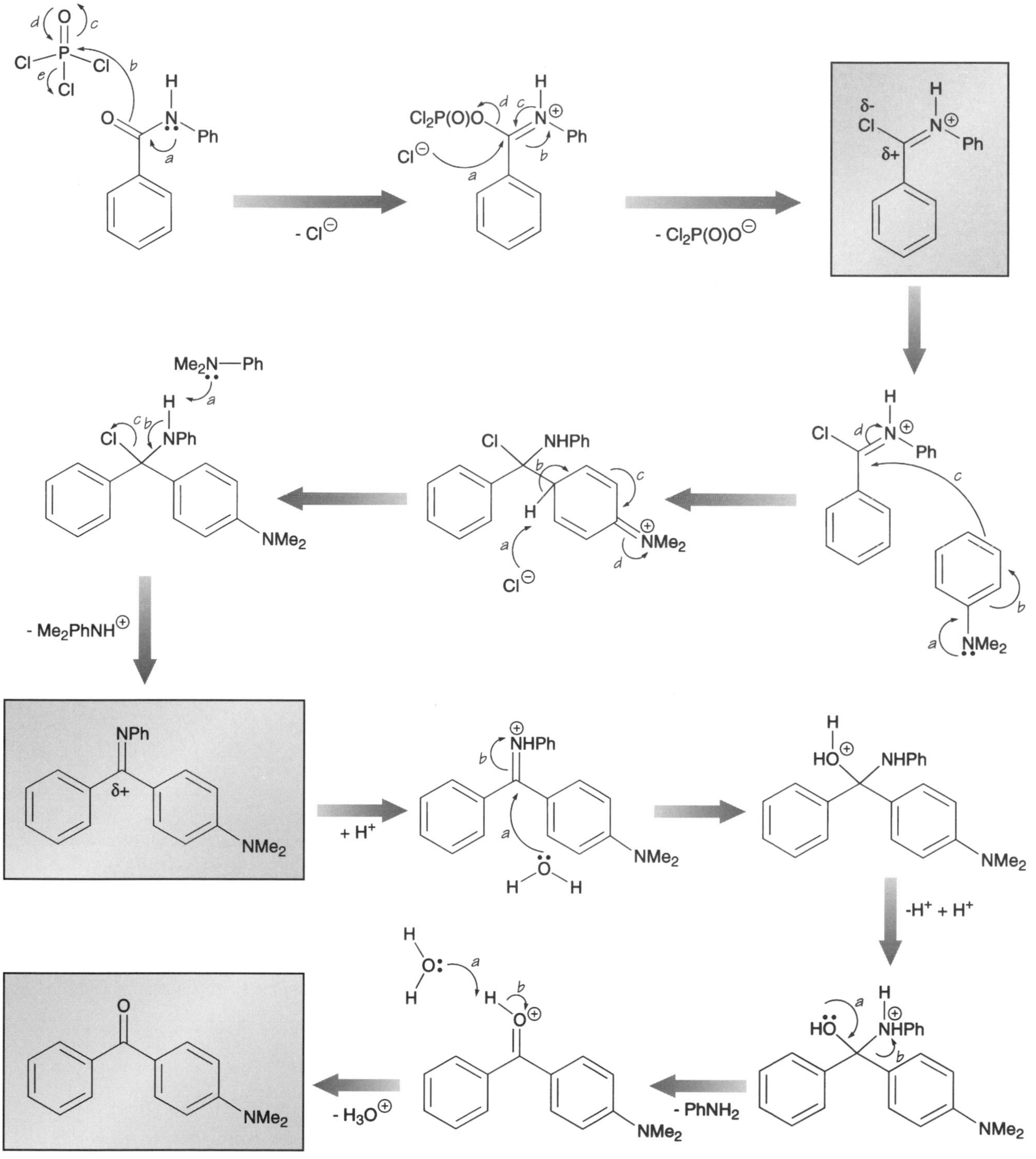

Summary: This is an example of Vilsmeier–Haack reaction.

Now try questions 5.12 and 5.13 at the end of this Chapter

5.4

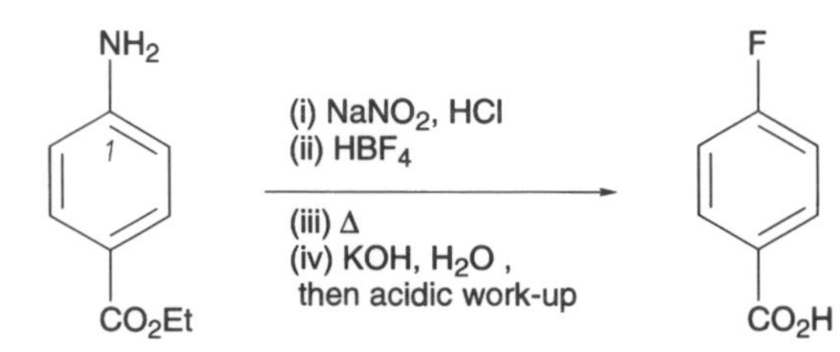

Label electrophilic/nucleophilic and acidic/basic sites of all reactants, and number identical atoms in the starting material and product

Esters and amines are nucleophiles (both oxygen and nitrogen possess lone pairs).

Identify the most reactive sites, if more than one exists

The amine is the most nucleophilic site, although it is deactivated a little by the ester substituent by a resonance interaction.

Recall the characteristic reactions of the most reactive functional groups and, by considering the reaction conditions, decide which is the most appropriate

The combination $NaNO_2$/HCl generates HONO (nitrous acid), which is in **equilibrium** with N_2O_3, a source of $O{=}N^+$. This reagent **diazotizes** the amine functional group.

Work through the mechanism leading to the intermediate product

Nucleophilic addition of the amine is followed by **tautomerization**, which permits **elimination** of H_2O to give the diazonium salt product; this is converted to the BF_4^- salt by treatment with fluoroboric acid.

Repeat the above four steps...

* The diazo group loses nitrogen gas on heating to give an aryl cation; this is intercepted by the available nucleophile, fluoride ion.
* Hydroxide is a good base and nucleophile, and causes alkaline hydrolysis (**addition–elimination** mechanism) of the ester group.

Recognize that this is not the final product, but is closely related to it

Acidic work-up gives the carboxylic acid product.

Write down the structure of the final product

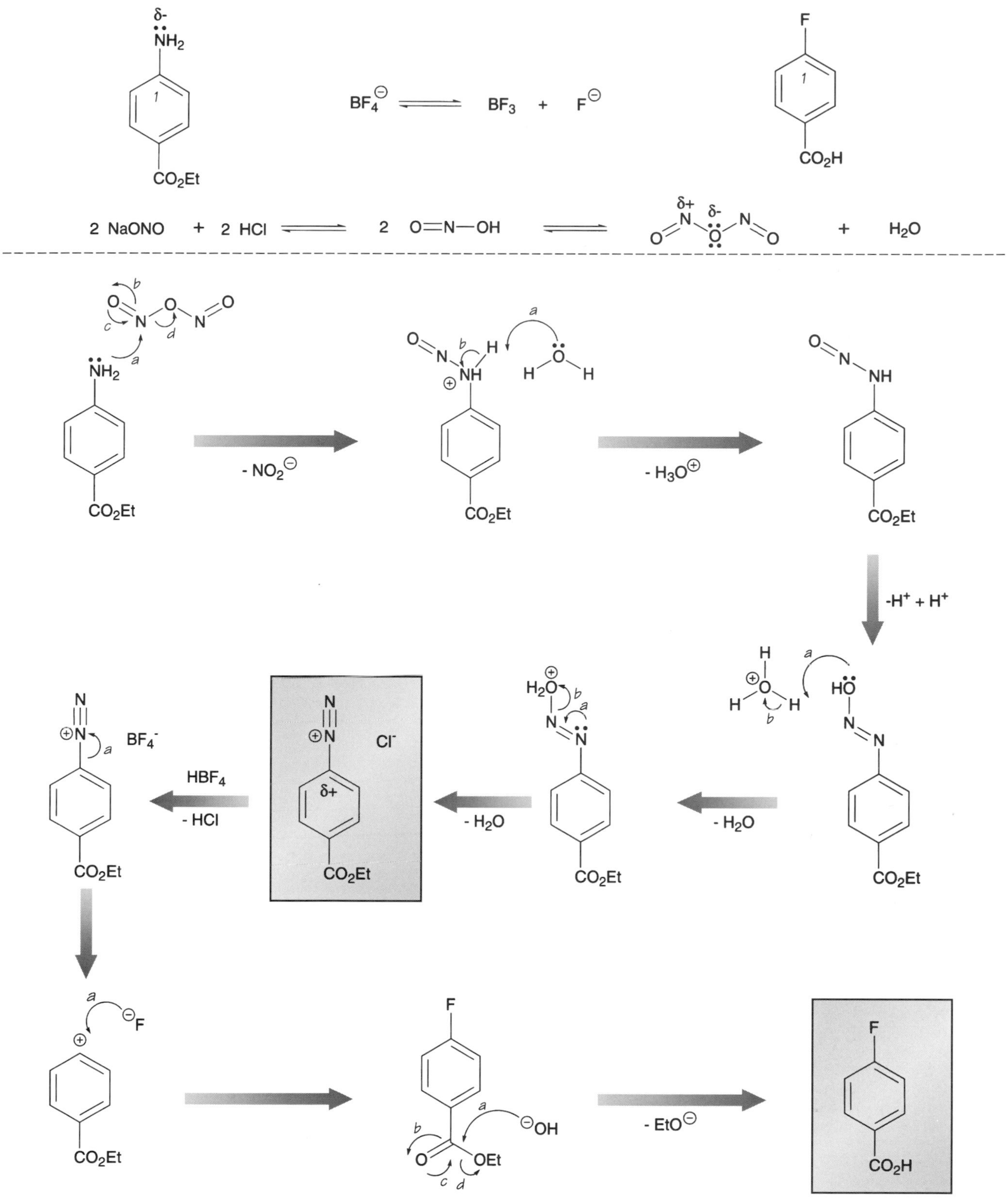

$$BF_4^{\ominus} \rightleftharpoons BF_3 + F^{\ominus}$$

$$2\ NaONO + 2\ HCl \rightleftharpoons 2\ O{=}N{-}OH \rightleftharpoons O{=}N^{\delta+}{-}O^{\delta-}{-}N{=}O + H_2O$$

$-NO_2^{\ominus}$

$-H_3O^{\oplus}$

$-H^+ + H^+$

$-H_2O$

$-H_2O$

HBF_4

$-HCl$

$-EtO^{\ominus}$

Summary: This is an example of the Schiemann reaction, involving the generation of a diazonium cation, followed by attack with F^-.

$$PhN_2^{\oplus} \longrightarrow PhF$$

Now try questions 5.14 and 5.15 at the end of this Chapter

5.5

O_2N, HO, 1, NO_2, O_2N → (i) PCl_5 (ii) 2 NH then basic work-up → N, O_2N, NO_2, O_2N

Label electrophilic/nucleophilic and acidic/basic sites of all reactants, and number identical atoms in the starting material and product

PCl_5 is an electrophilic chlorinating agent (there are five electron-withdrawing chlorines and each is a good leaving group). Phenols are nucleophilic at oxygen.

Identify the most reactive sites, if more than one exists

The oxygen of the phenol is the most nucleophilic site.

Recall the characteristic reactions of the most reactive functional groups and, by considering the reaction conditions, decide which is the most appropriate

Alcohols react with phosphorus halides to give the corresponding alkyl chloride.

Work through the mechanism leading to the intermediate product

Initial attack by the hydroxyl group on phosphorus generates a good **leaving group** at C-1; a sequence of **addition–elimination** steps then gives the chloride product, giving overall substitution of chloride by hydroxide.

Repeat the above four steps...

Amines are nucleophilic at nitrogen. The aromatic chloride is highly electron-deficient and carries a good **leaving group**, so attack by the nucleophilic piperidine then generates the corresponding product by an **addition– elimination** reaction.

Recognize that this is not the final product, but is closely related to it

Deprotonation on aqueous work-up gives the final product, since the ammonium product is very acidic.

Write down the structure of the final product

Summary: This is an example of nucleophilic aromatic substitution (S_NAr).

Now try questions 5.16 and 5.17 at the end of this Chapter

5.6

(i) n-BuLi, Et_2O
(ii) Me_2CuLi then $H_2C{=}CHCH_2Br$
(iii) HCl, H_2O

Label electrophilic/nucleophilic and acidic/basic sites of all reactants, and number identical atoms in the starting material and product

n-BuLi is a strong base (pK_a of butane = 50).
The aromatic protons α to the chloro substituent are weakly acidic.

Identify the most reactive sites, if more than one exists

The proton at C-2 is the more acidic, since the corresponding lithiated species is stabilized by chelation to the adjacent nitrogen atom.

Identify the characteristic reactions of the most reactive functional groups and, by considering the reaction conditions, decide which is the most appropriate

Aromatic halides readily react with strong bases to give a **benzyne** intermediate.

Work through the mechanism leading to the intermediate product

β-Elimination gives the benzyne intermediate directly.

Repeat the above four steps...

* Benzyne is electrophilic, and is readily attacked by nucleophiles, in this case dimethylcuprate. The resulting anion is trapped by allyl bromide in a nucleophilic substitution reaction.
* The oxazoline intermediate is hydrolysed to the corresponding carboxylic acid product.

Recognize that this is not the final product, but is closely related to it

Not needed here.

Write down the structure of the final product

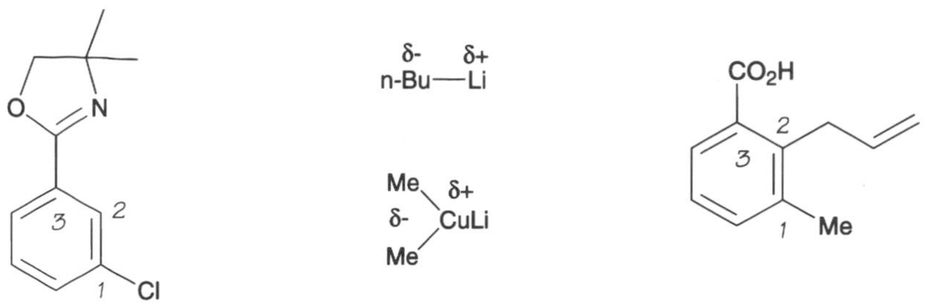

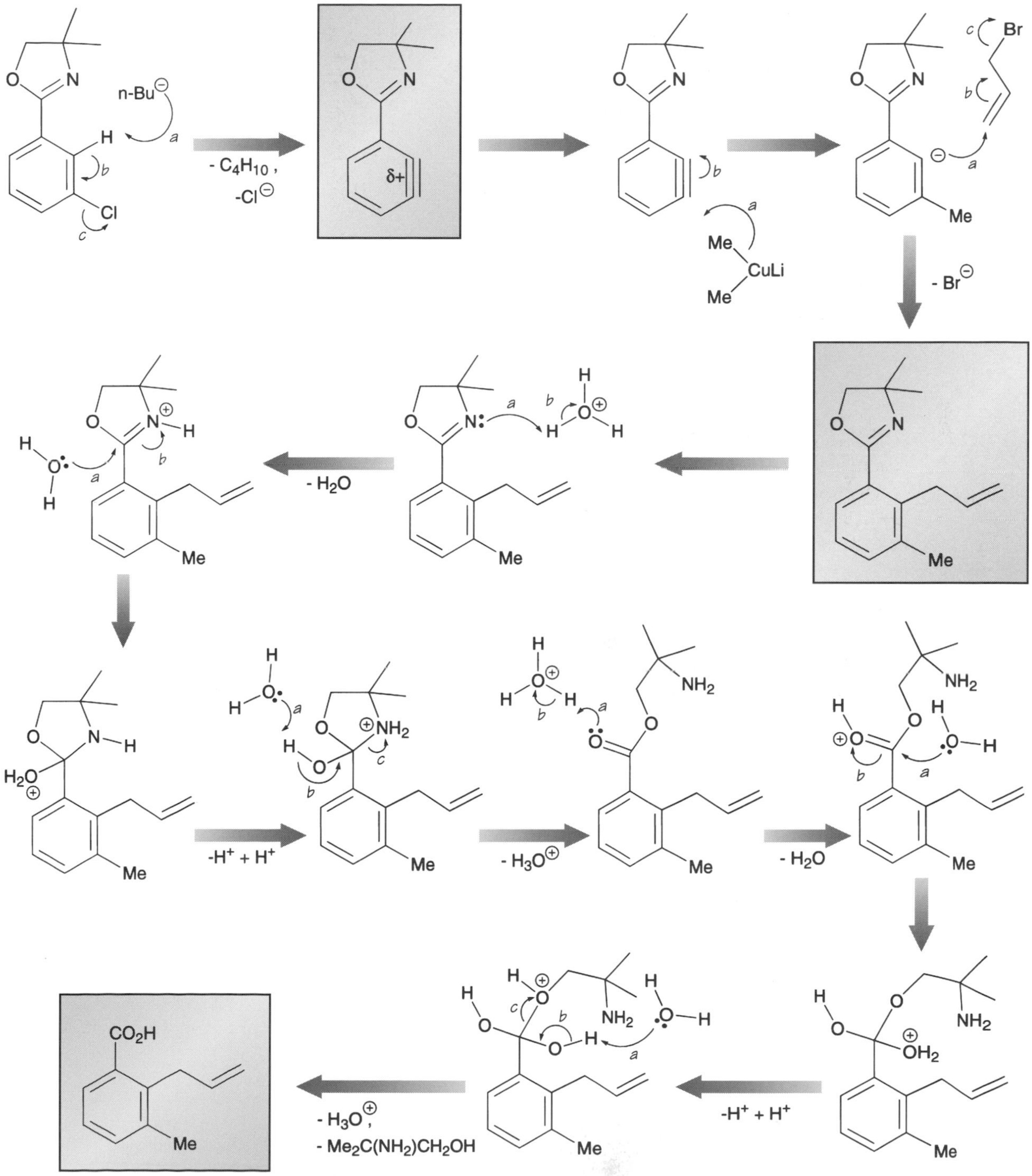

Summary: This is an example involving a benzyne intermediate.

Now try questions 5.18 and 5.19 at the end of this Chapter

5.7

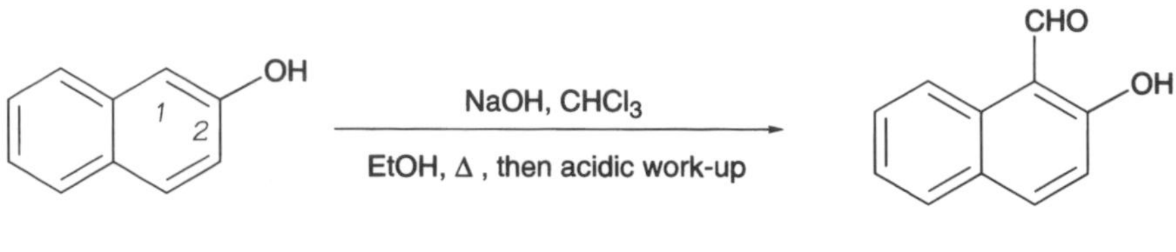

Label electrophilic/nucleophilic and acidic/basic sites of all reactants, and number identical atoms in the starting material and product

2-Naphthol is a highly nucleophilic aromatic compound, by virtue of the **phenolic** oxygen. NaOH is a strong base, and $CHCl_3$ possesses an acidic hydrogen.

Identify the most reactive sites, if more than one exists

C-1 of 2-naphthol is especially activated (*o*-to the hydroxy group) to electrophilic attack.

Recall the characteristic reactions of the most reactive functional groups and, by considering the reaction conditions, decide which is the most appropriate

NaOH deprotonates $CHCl_3$, and will give **dichlorocarbene** by an **α-elimination** reaction. This intermediate is very electrophilic. Under these conditions, naphthol is also deprotonated to the alkoxide.

Work through the mechanism leading to the intermediate product

Electrophilic aromatic substitution at C-1 of 2-naphthol with dichlorocarbene gives the dichloromethyl product.

Repeat the above four steps...

Elimination o f chloride generates an enone which is intercepted by hydroxide. Collapse of the intermediate so formed generates the aldehyde product.

Recognize that this is not the final product, but is closely related to it

Acidic work-up reprotonates the phenoxide anion, giving the product.

Write down the structure of the final product

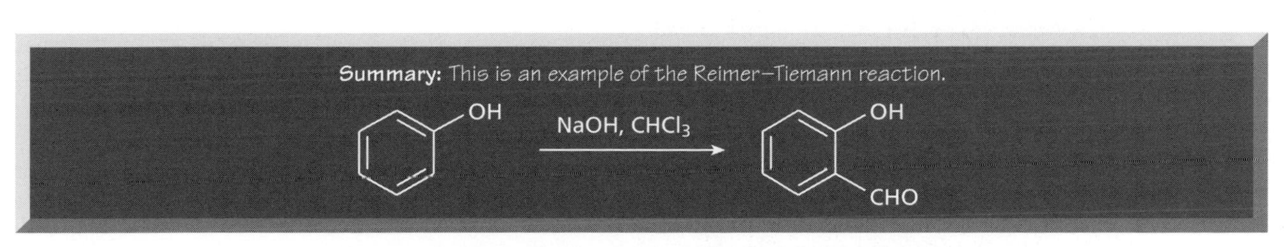
Summary: This is an example of the Reimer–Tiemann reaction.
OH
NaOH, CHCl3
OH
CHO

Now try questions 5.20 and 5.21 at the end of this Chapter

Supplementary questions

5.8

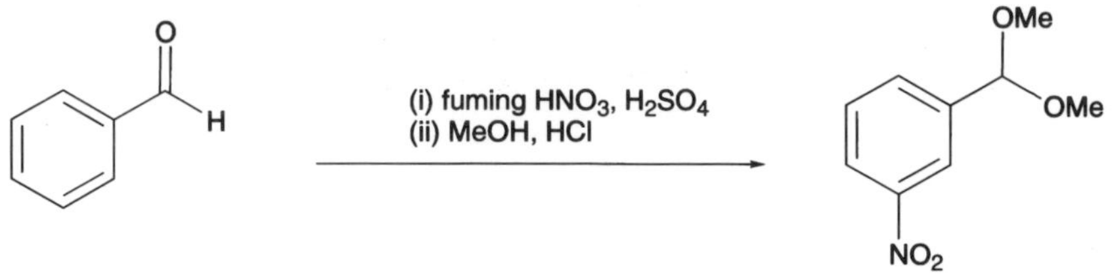
O
H
(i) fuming HNO3, H2SO4
(ii) MeOH, HCl
OMe
OMe
NO2

5.9

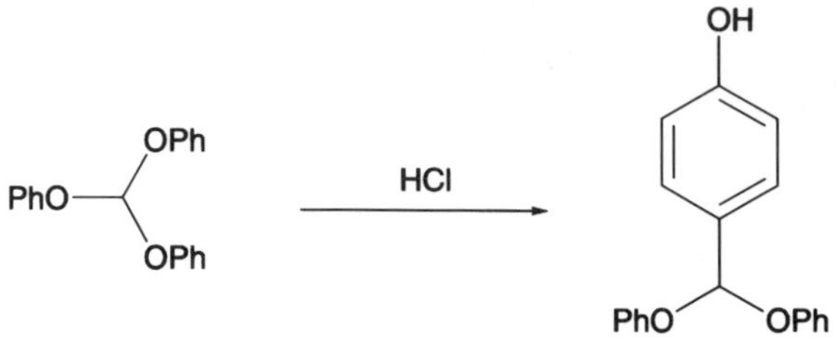
OPh
PhO
OPh
HCl
OH
PhO
OPh

5.10

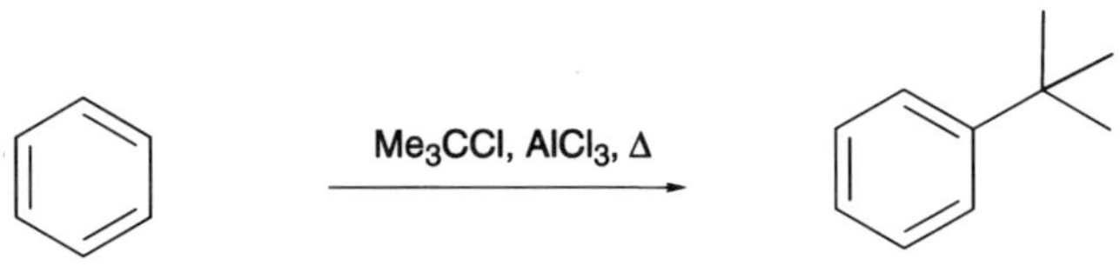
Me3CCl, AlCl3, Δ

5.11

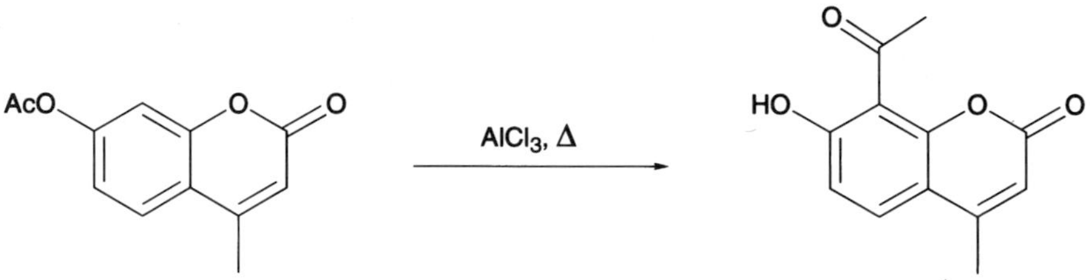
AcO
O
O
AlCl3, Δ
O
HO
O
O

5.12

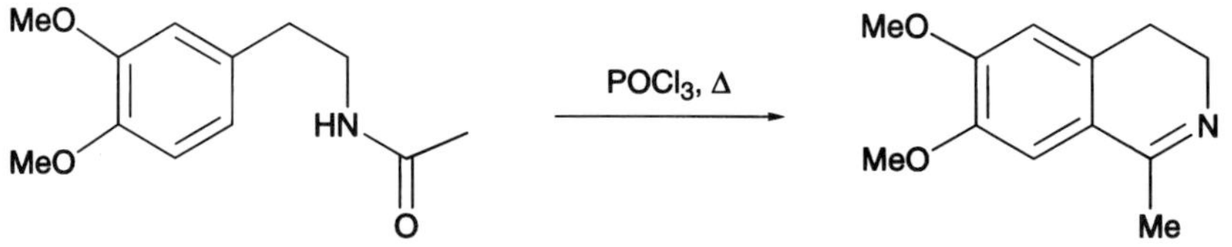
MeO
MeO
HN
O
POCl3, Δ
MeO
MeO
N
Me

5.13

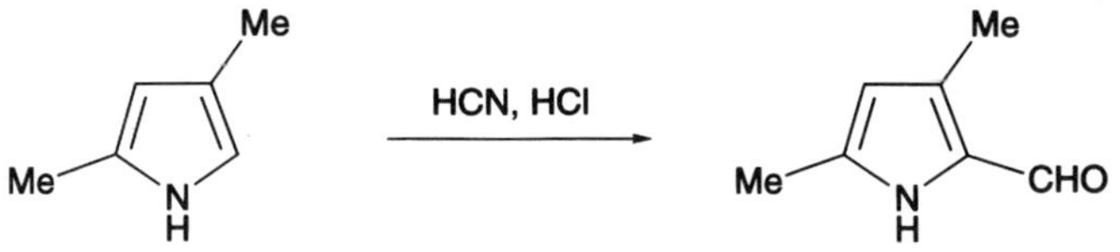
Me
Me
N
H
HCN, HCl
Me
Me
N
H
CHO

5.14

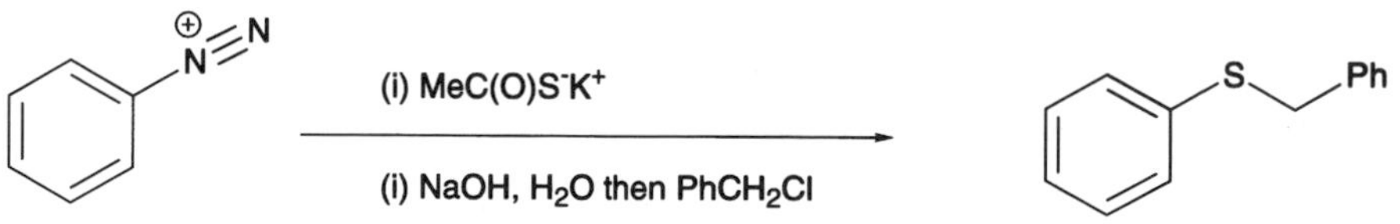

(i) $MeC(O)S^-K^+$

(i) NaOH, H_2O then $PhCH_2Cl$

5.15

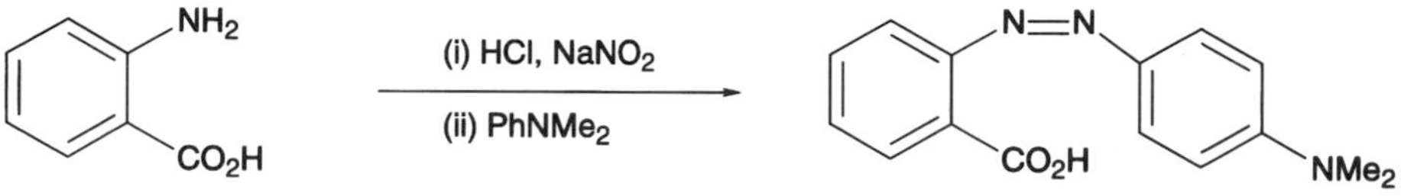

NH_2

CO_2H

(i) HCl, $NaNO_2$

(ii) $PhNMe_2$

N=N

CO_2H

NMe_2

5.16

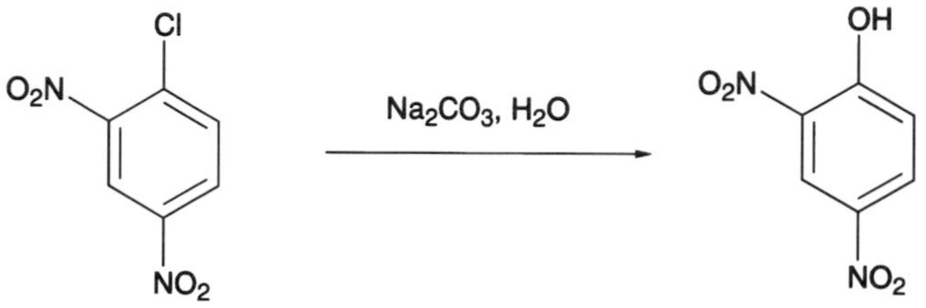

Cl

O_2N

NO_2

Na_2CO_3, H_2O

OH

O_2N

NO_2

5.17

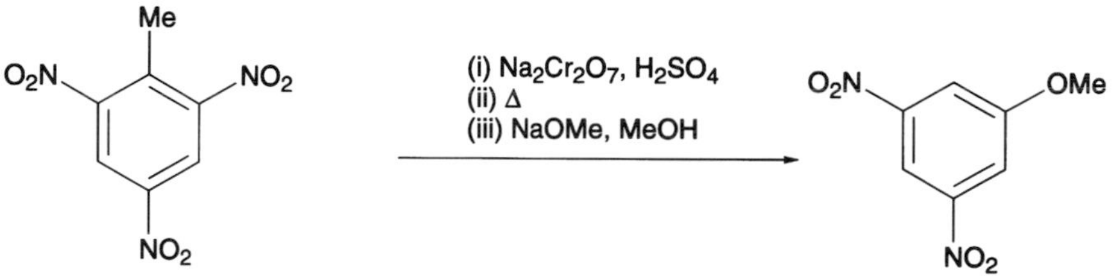

Me

O_2N

NO_2

NO_2

(i) $Na_2Cr_2O_7$, H_2SO_4
(ii) Δ
(iii) NaOMe, MeOH

O_2N

OMe

NO_2

5.18

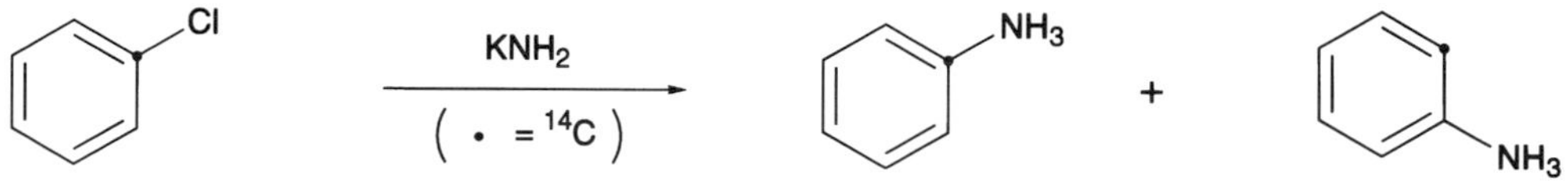

Cl

KNH_2

(• = ^{14}C)

NH_3

+

NH_3

5.19

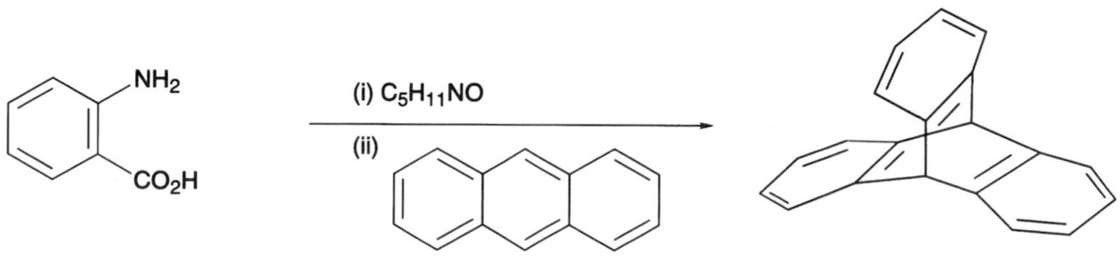

NH_2

CO_2H

(i) $C_5H_{11}NO$

(ii)

5.20

2-phenylindole $\xrightarrow{CHCl_3,\ NaOH}$ 2-phenylindole-3-carbaldehyde (CHO, Ph, N–H)

5.21

MeO, MeO, Me $\xrightarrow[\text{(ii) HCl, } H_2O \text{ (iii) } NaBH_4 \text{, then acidic work-up}]{\text{(i) } Cl_2CHOMe,\ TiCl_4}$ MeO, MeO, Me, CH_2OH

6 Rearrangements

Rearrangements most commonly occur in compounds which possess an electron-releasing group (X), a migrating group (R) and a leaving group (L); the ability for correct orbital overlap is critical for the reaction pathway to occur, and there is a clear preference for an *anti-* relationship between migrating and leaving groups. Aryl and alkyl shifts are generally preferred over hydrogen shifts, with aryl more likely than alkyl. The reaction can be represented in its broadest form as follows:

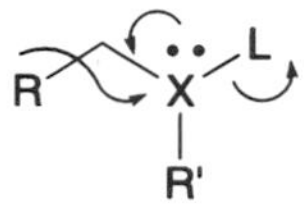

There are several types:

1,2- Alkyl or aryl shifts to carbon

The stability of carbocations, and the relative migrations of substituents, is very important and this determines the course of the reaction.

Examples are:

- Wagner–Meerwein
- Pinacol–pinacolone
- Benzil–benzilic acid
- Favorskii
- Tiffenau–Demyanov
- Dienone–phenol
- Wolff (Arndt–Eistert reaction)

1,2- Alkyl or aryl shifts to nitrogen or oxygen

Examples are:

- Beckmann (reaction of oximes with PCl_5)
- Hofmann (reaction of a primary amide with alkaline bromine solution)
- Schmidt (reaction of carbonyl compounds with HN_3)
- Lossen (reaction of an *O*-acyl hydroxamic acid under basic conditions)
- Curtius (pyrolysis of acyl azides)
- Baeyer–Villiger (reaction of ketones with peracids)
- Hydroperoxide (reaction of hydroperoxides under acidic conditions)

Hydride transfer reactions

Examples are:

- Cannizzarro (reaction of aromatic and some aliphatic aldehydes under alkaline conditions)
- Meerwein–Ponndorf–Verley reduction and Oppenauer oxidation (disproportionation of a ketone/isopropyl alcohol or alcohol/acetone mixture, respectively, in the presence of aluminium isopropoxide)

6.1

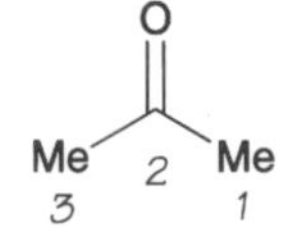

(i) Mg , then acidic work-up

(iii) conc H_2SO_4, Δ
(iv) $NaBH_4$, then acidic work-up

HO Me Me Me Me

Label electrophilic/nucleophilic and acidic/basic sites of all reactants, and number identical atoms in the starting material and product

The carbonyl group is polarized by the **electronegativity** difference between carbon and oxygen, making the carbon electrophilic.
Magnesium is a reducing metal, and can donate two electrons, in two separate single-electron processes.

Identify the most reactive sites, if more than one exists

The ketone is the only reactive site.

Recall the characteristic reactions of the most reactive functional groups and, by considering the reaction conditions, decide which is the most appropriate

Reducing metals (such as Mg) can couple ketones to give 1,2-diols.

Work through the mechanism leading to the intermediate product

Magnesium donates a single electron to the oxygen of the C=O bond, to form a stable O–Mg bond, and a carbon-centred *radical*; this occurs twice to give a magnesium **bis(alkoxide)**. The radicals then **dimerize**, and the **bis(alkoxide)** product is protonated on work-up to give a diol.

Repeat the above four steps...

* Alcohols are basic and nucleophilic. Treatment with strong acid (H_2SO_4) protonates one of the hydroxyl groups, and this induces a **rearrangement**: the lone pair of the other hydroxyl pushes in, forces a 1,2-methyl group migration, and is followed by departure of the **leaving group** (H_2O).
* $NaBH_4$ is a source of (nucleophilic) hydride. The carbonyl group is polarized by the **electronegativity** difference between carbon and oxygen, making the carbon electrophilic. **Nucleophilic addition** of hydride generates an alkoxide anion.

Recognize that this is not the final product, but is closely related to it

Protonation on work-up generates the alcohol product.

Write down the structure of the final product

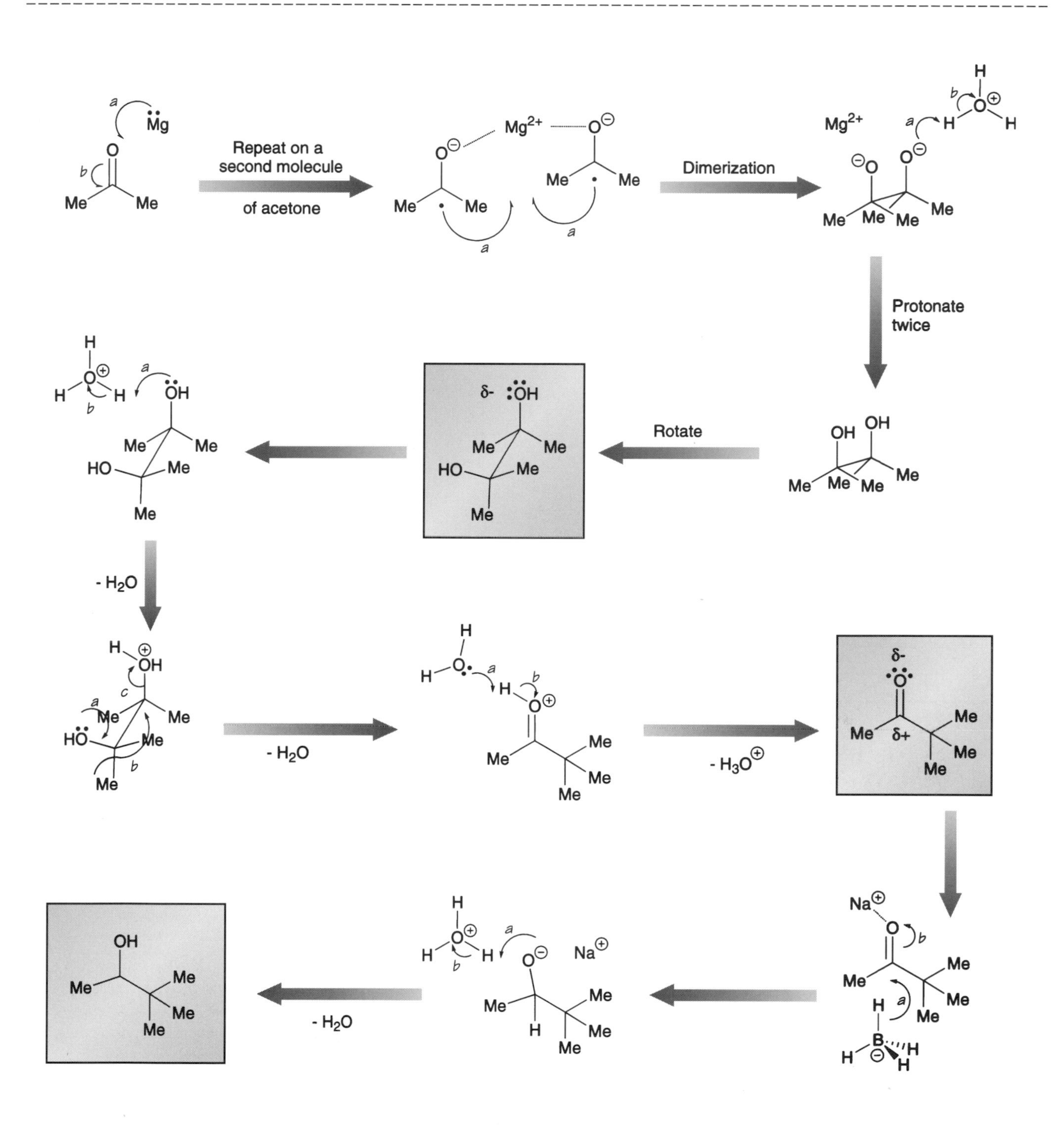

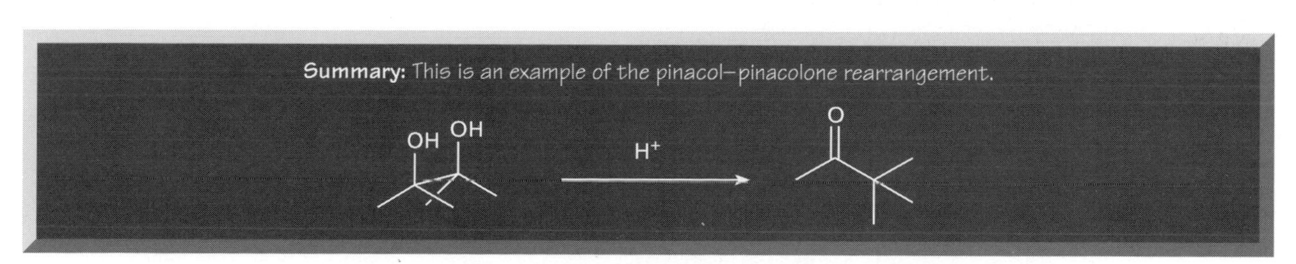

Summary: This is an example of the pinacol–pinacolone rearrangement.

Now try questions 6.8 and 6.9 at the end of this Chapter

6.2

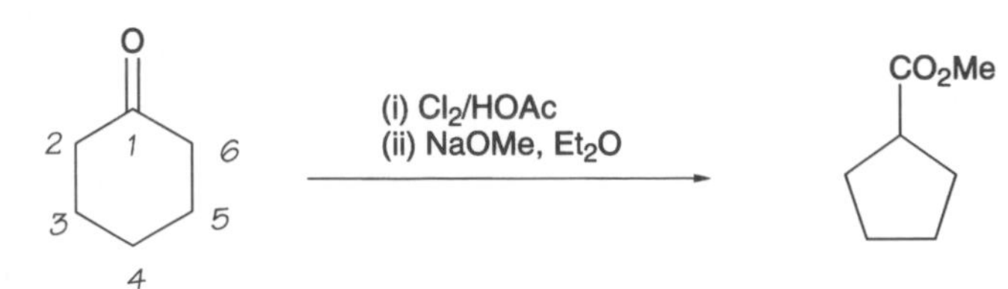

Label electrophilic/nucleophilic and acidic/basic sites of all reactants, and number identical atoms in the starting material and product

The carbonyl group is polarized by the **electronegativity** difference between carbon and oxygen, making the carbon electrophilic.
Chlorine is an electrophile, with a weak Cl–Cl bond.

Identify the most reactive sites, if more than one exists

The ketone is the only reactive group.

Recall the characteristic reactions of the most reactive functional groups and, by considering the reaction conditions, decide which is the most appropriate

Ketones are readily halogenated at the α position by reaction of their corresponding enol (easily generated under acidic conditions by **tautomerization**).

Work through the mechanism leading to the intermediate product

The α-position of the enol of cyclohexanone (here C-6) attacks chlorine, to give an α-chloroketone.

Repeat the above four steps...

* Sodium methoxide is a strong base. Deprotonation at C-2 gives an enolate, and this undergoes an intramolecular **nucleophilic substitution** at the carbon bearing the chlorine, to generate a cyclopropanone.
* Cyclopropanones are particularly susceptible to **nucleophilic addition** due to ring strain. Methoxide is a good base and nucleophile, and undergoes an **addition–elimination** reaction at the carbonyl with concomitant ring opening.

Recognize that this is not the final product, but is closely related to it

Protonation of the alkyl anion by methanol gives the product.

Write down the structure of the final product

O
δ+
1
2 6
3 5
4

:Cl—Cl:

1 CO_2Me
2
3 6
4 5

AcO—H
c
b
O
H
a
H

- AcO⊖

H
O
a
b
c
Cl—Cl
δ+

- Cl⊖

H
O
δ-
Cl
H
δ+
H

MeO⊖ a
O
c
H b
Cl
H
H

- MeOH

O⊖
Cl
H

O
Cl
b
⊖
a
H

- Cl⊖

δ-
O
δ+

b O a ⊖OMe

a
⊖
O
OMe
b

MeO
O
⊖
a b
H O Me

- MeO⊖

MeO
O

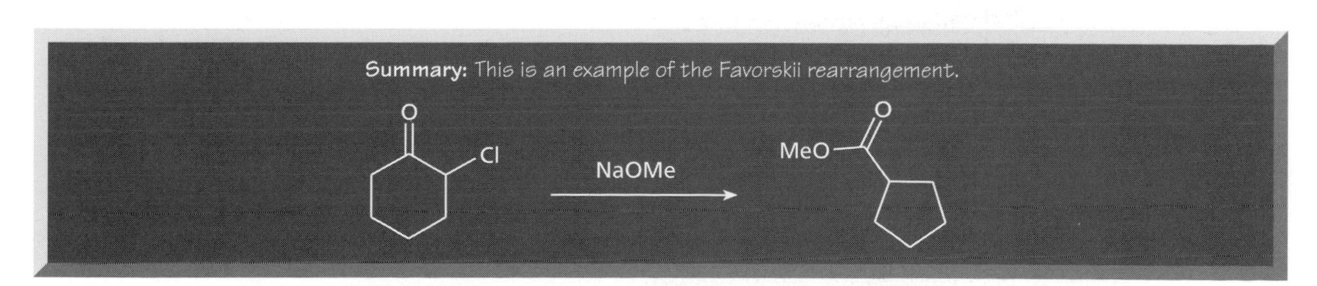

Summary: This is an example of the Favorskii rearrangement.

Now try questions 6.10 and 6.11 at the end of this Chapter

6.3

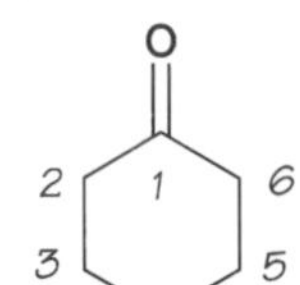

(i) NH_2OH, H_2O, EtOH
(ii) H_2SO_4, Δ
(iii) NaOH, H_2O

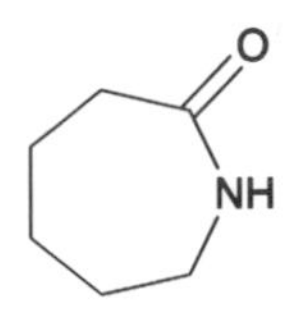

Label electrophilic/nucleophilic and acidic/basic sites of all reactants, and number identical atoms in the starting material and product

The carbonyl group is polarized by the **electronegativity** difference between carbon and oxygen, making the carbon electrophilic.
Hydroxylamine is a very good nucleophile (two adjacent heteroatoms).

Identify the most reactive sites, if more than one exists

The ketone is the only reactive group

Recall the characteristic reactions of the most reactive functional groups and, by considering the reaction conditions, decide which is the most appropriate

Hydroxylamine and ketones react to give **oximes**.

Work through the mechanism leading to the intermediate product

Addition–elimination of hydroxylamine to the ketone, with expulsion of water, gives an oxime.

Repeat the above four steps...

H_2SO_4 is a strong acid and fully ionized, and oximes are basic on oxygen.
Protonation of the **oxime** generates a **leaving group**; the adjacent alkyl group migrates with departure of the leaving group to give a vinylic **carbocation**. This is intercepted by water, to give the product, an amide in its enolic form.

Recognize that this is not the final product, but is closely related to it

Tautomerization to the ketone form gives the amide product.

Write down the structure of the final product

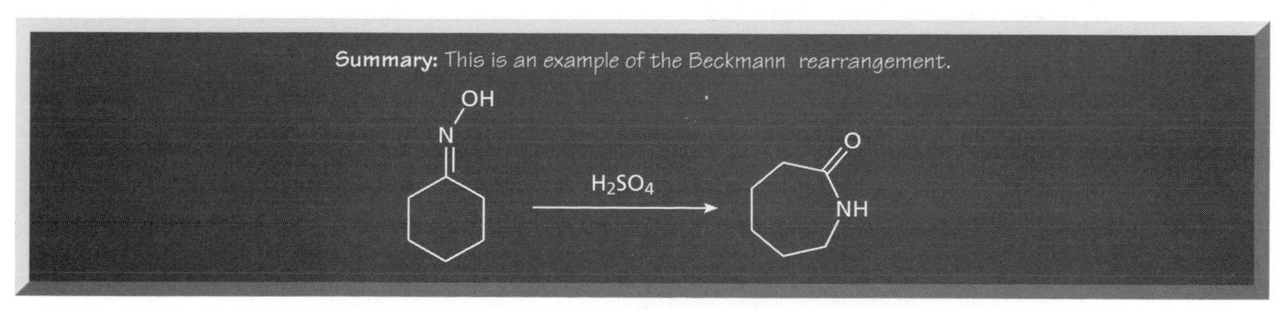
Summary: This is an example of the Beckmann rearrangement.
OH
N
H2SO4
O
NH

Now try questions 6.12 and 6.13 at the end of this Chapter

6.4

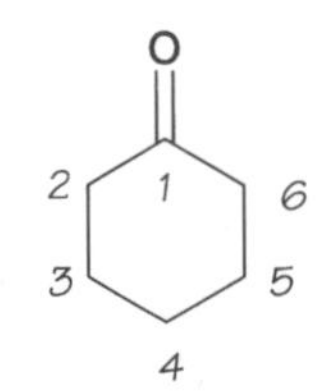

(i) CH_3NO_2, NaOEt
(ii) H_2, Raney nickel, HOAc

(iii) $NaNO_2$, HOAc, H_2O, 0°C

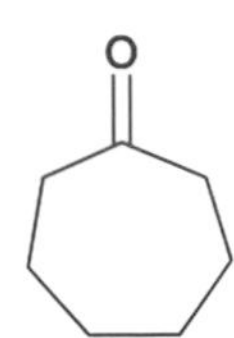

Label electrophilic/nucleophilic and acidic/basic sites of all reactants, and number identical atoms in the starting material and product

The carbonyl group is polarized by the **electronegativity** difference between carbon and oxygen, making the carbon electrophilic.
Nitromethane is relatively acidic (pK_a = 10), easily deprotonated by sodium ethoxide, to give a **nitronate** anion.

Identify the most reactive sites, if more than one exists

The ketone is the only reactive group.

Recall the characteristic reactions of the most reactive functional groups and, by considering the reaction conditions, decide which is the most appropriate

The carbon of the ketone is susceptible to **nucleophilic addition**.

Work through the mechanism leading to the intermediate product

Nucleophilic addition of the nitronate anion to the carbonyl gives an alkoxide; protonation on work-up gives the alcohol.

Repeat the above four steps...

* Raney nickel is a good reducing agent, capable of converting a nitro group to an amine (the mechanism is not well understood).
* **Diazotization** with nitrous acid generates a very good leaving group (nitrogen). The oxygen lone pair pushes in, forces a **1,2-alkyl shift** with loss of nitrogen to give the intermediate product, which is the protonated form of cycloheptanone.

Recognize that this is not the final product, but is closely related to it

Deprotonation to the keto form gives the product, cycloheptanone.

Write down the structure of the final product

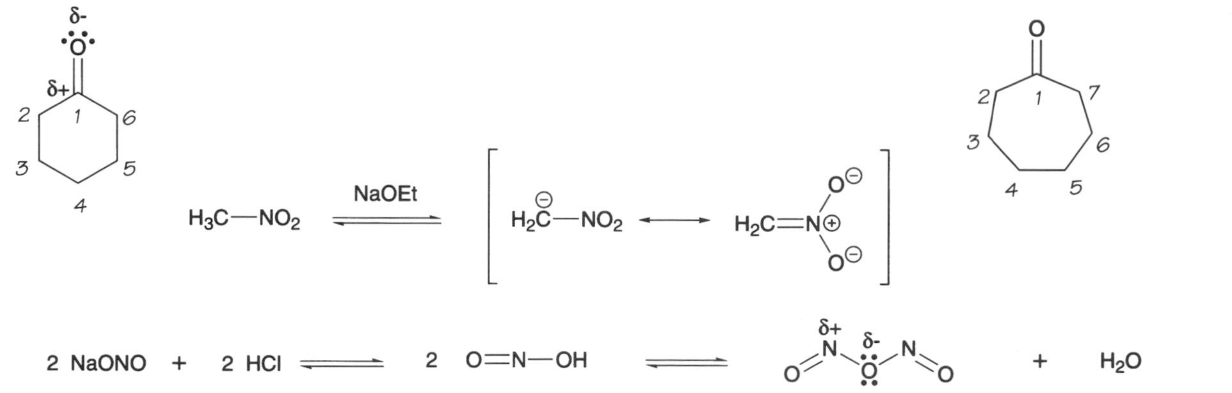

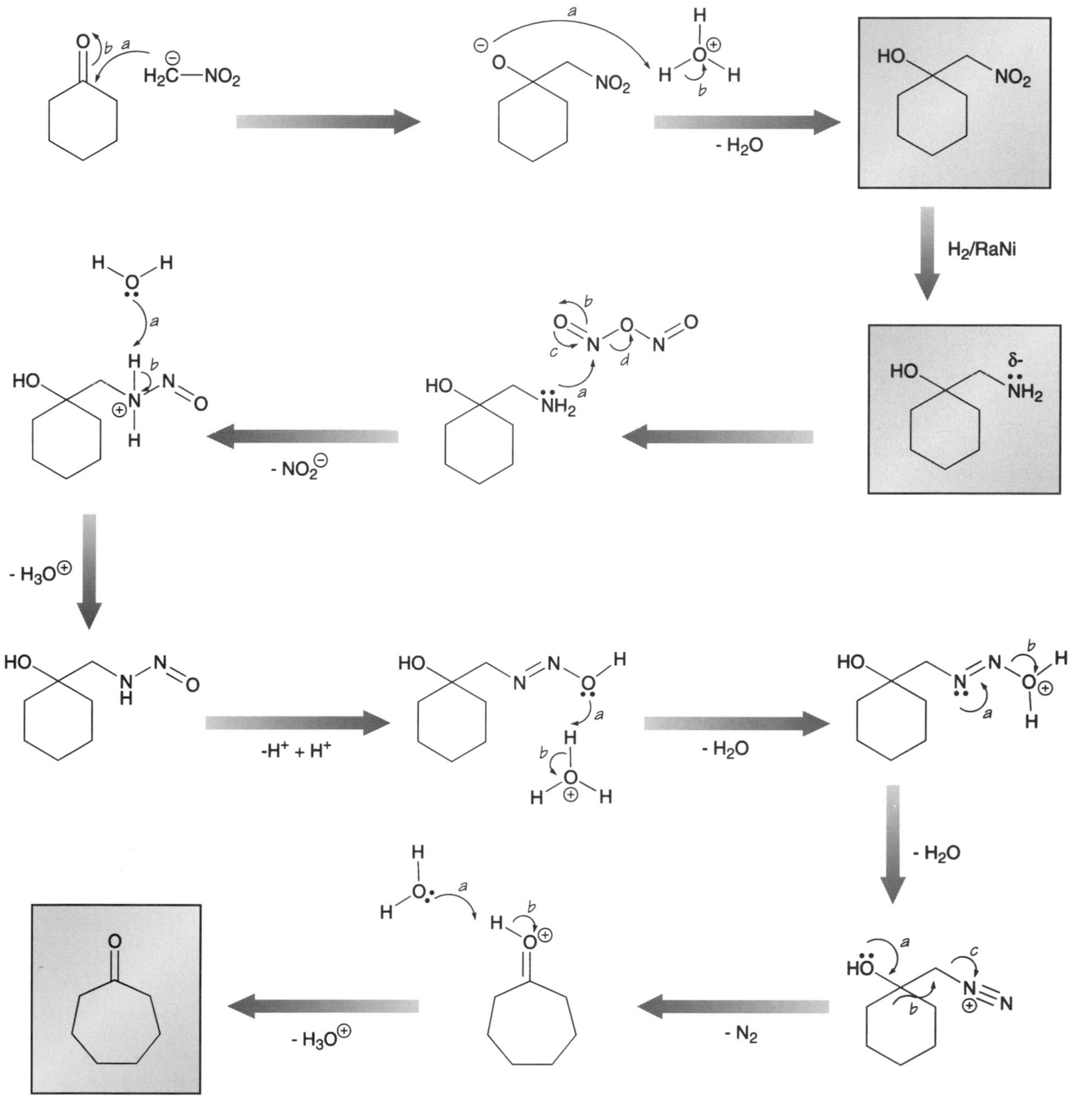

δ- δ+ 1 2 3 4 5 6

$H_3C—NO_2$ ⇌ (NaOEt) [$H_2\overset{\ominus}{C}—NO_2$ ⟷ $H_2C{=}N^{\oplus}(O^{\ominus})_2$]

1 2 3 4 5 6 7

2 NaONO + 2 HCl ⇌ 2 O=N—OH ⇌ $O{=}N{-}O{-}N{=}O$ (δ+, δ-) + H_2O

$H_2\overset{\ominus}{C}—NO_2$ → a, b → NO_2 → $-H_2O$ → HO, NO_2

H_2/RaNi

HO, NH_2 (δ-)

HO, NH_2 → $-NO_2^{\ominus}$ → HO, $\oplus$N, H, N=O

$-H_3O^{\oplus}$

HO, N–N=O, H → $-H^+ + H^+$ → HO, N=N–OH → $-H_2O$ → HO, N=N–$O^{\oplus}H_2$

$-H_2O$

HO, $N^{\oplus}{\equiv}N$ → $-N_2$ → $H–O^{\oplus}$= (cycloheptanone) → $-H_3O^{\oplus}$ → cycloheptanone

Summary: This is an example of the Tiffenau–Demjanov rearrangement.

Now try questions 6.14 and 6.15 at the end of this Chapter

6.5

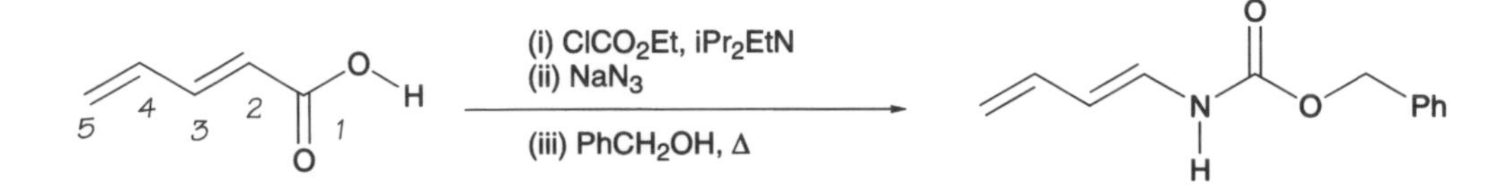

Label electrophilic/nucleophilic and acidic/basic sites of all reactants, and number identical atoms in the starting material and product

A carboxylic acid is electrophilic at the carbonyl carbon and nucleophilic at the oxygen atoms.
A chloroformate is electrophilic at the carbonyl carbon and carries two leaving groups.

Identify the most reactive sites, if more than one exists

The chloroformate is the most reactive electrophile due to the highly electron-withdrawing chlorine. This means that the carboxylic acid must act as a nucleophile.

Recall the characteristic reactions of the most reactive functional groups and, by considering the reaction conditions, decide which is the most appropriate

Chloroformates are highly reactive to nucleophilic **addition–elimination**.

Work through the mechanism leading to the intermediate product

Addition-elimination of the nucleophilic carboxylic acid to the chloroformate gives a mixed anhydride after expulsion of chloride.

Repeat the above four steps...

* A **mixed anhydride** is very susceptible to nucleophilic attack; azide is an excellent nucleophile, and adds with loss of CO_2 and ethoxide.
* An acyl azide can rearrange; donation of the nitrogen lone pair, **1,2-alkyl shift** and loss of nitrogen gas occurs to give a highly reactive **isocyanate**; addition of benzyl alcohol to the carbonyl then occurs.

Recognize that this is not the final product, but is closely related to it

Proton shift then gives the benzyloxy carbamate product.

Write down the structure of the final product

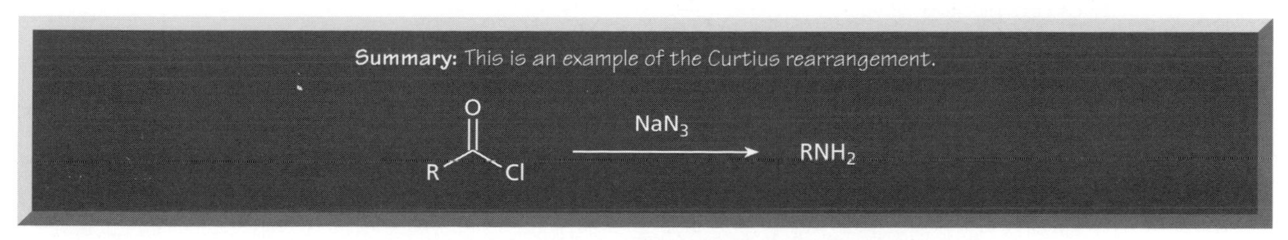
Summary: This is an example of the Curtius rearrangement.
O
R
Cl
NaN_3
RNH_2

Now try questions 6.16 and 6.17 at the end of this Chapter

6.6

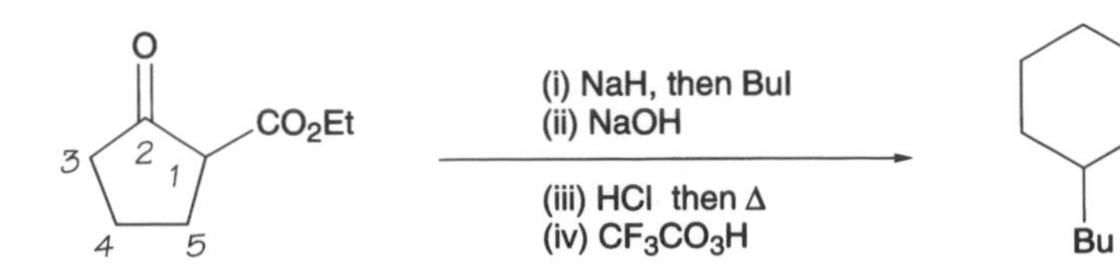

Label electrophilic/nucleophilic and acidic/basic sites of all reactants, and number identical atoms in the starting material and product

The α-protons of β-dicarbonyl compounds are highly acidic (pK_a about 13). Sodium hydride is a strong base. Iodobutane is a good electrophile (iodide is a good leaving group and its electronegativity polarizes the carbon–iodine bond).

Identify the most reactive sites, if more than one exists

The α-position of the β-dicarbonyl enolate is the most nucleophilic (here labelled C-1).

Recall the characteristic reactions of the most reactive functional groups and, by considering the reaction conditions, decide which is the most appropriate

β-Dicarbonyl compounds are readily deprotonated and then alkylated.

Work through the mechanism leading to the intermediate product

Deprotonation of the β-dicarbonyl compound is followed by **nucleophilic substitution** of iodobutane.

Repeat the above four steps...

* Ester **hydrolysis** mediated by sodium hydroxide, followed by decarboxylation, gives an enol intermediate, which then **tautomerizes** to the ketone product.
* Addition of the hydroxyl group of trifluoroperacetic acid to the ketone gives a tetrahedral intermediate.
* Donation of the lone pair of the oxygen, **1,2-migration** of one of the alkyl groups and loss of the carboxylate anion then give the product lactone.

Recognize that this is not the final product, but is closely related to it

Not needed here.

Write down the structure of the final product

Summary: This is an example of the Baeyer–Villiger reaction.

$$RCOR' \xrightarrow{R''CO_3H} RCO_2R'$$

Now try questions 6.18 and 6.19 at the end of this Chapter

Supplementary questions

6.8

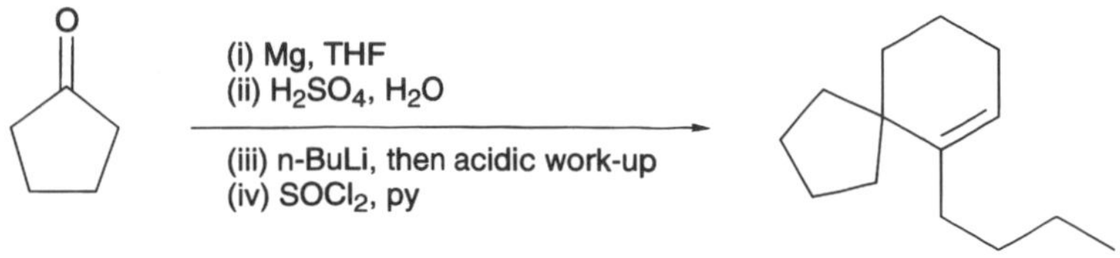

(i) Mg, THF
(ii) H_2SO_4, H_2O

(iii) n-BuLi, then acidic work-up
(iv) $SOCl_2$, py

6.9

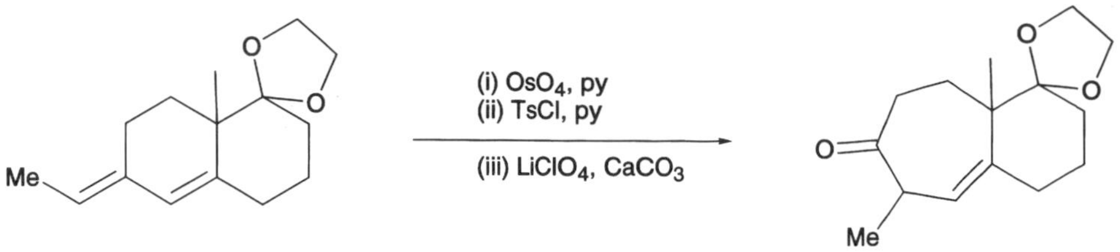

Me

(i) OsO_4, py
(ii) TsCl, py

(iii) $LiClO_4$, $CaCO_3$

Me

6.10

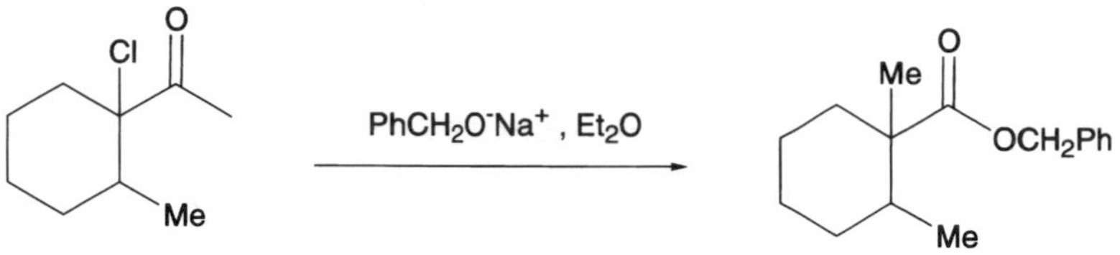

Cl O

Me

$PhCH_2O^-Na^+$, Et_2O

Me O

OCH_2Ph

Me

6.11

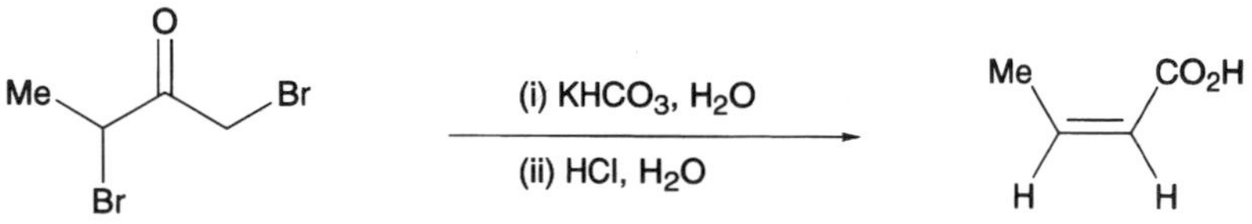

Me O Br

Br

(i) $KHCO_3$, H_2O

(ii) HCl, H_2O

Me CO_2H

H H

6.12

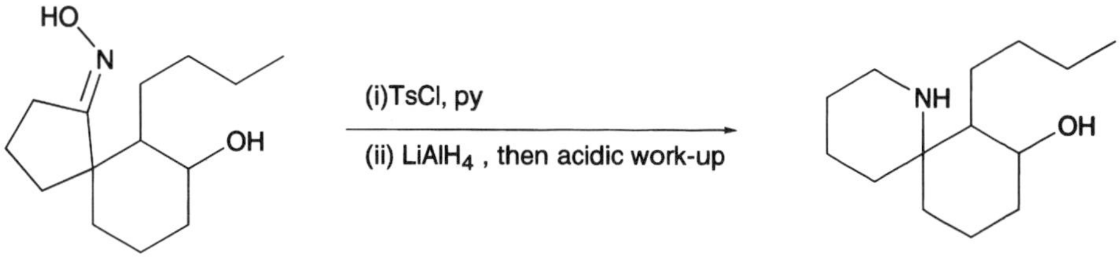

HO N

OH

(i)TsCl, py

(ii) $LiAlH_4$, then acidic work-up

NH OH

6.13

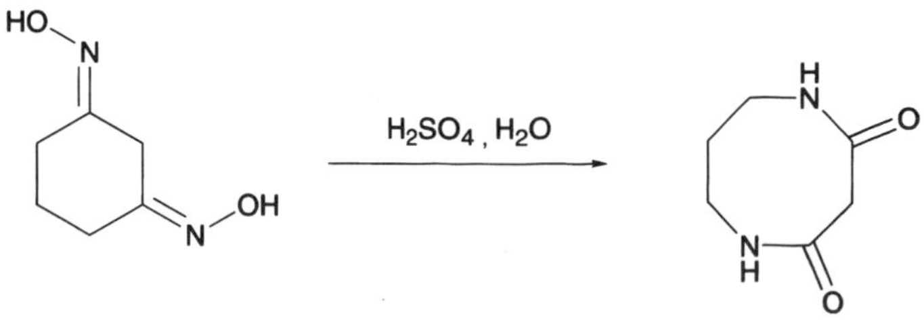

HO N

N OH

H_2SO_4 , H_2O

H N O

N H O

6.14

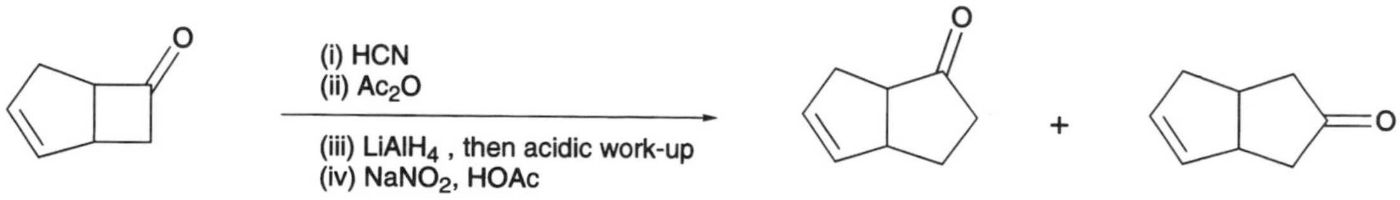

(i) HCN
(ii) Ac_2O
(iii) $LiAlH_4$, then acidic work-up
(iv) $NaNO_2$, HOAc

+

6.15

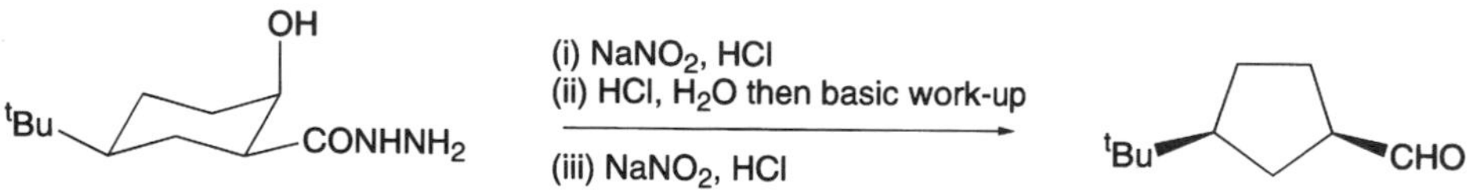

OH

^tBu CONHNH$_2$

(i) $NaNO_2$, HCl
(ii) HCl, H_2O then basic work-up
(iii) $NaNO_2$, HCl

^tBu CHO

6.16

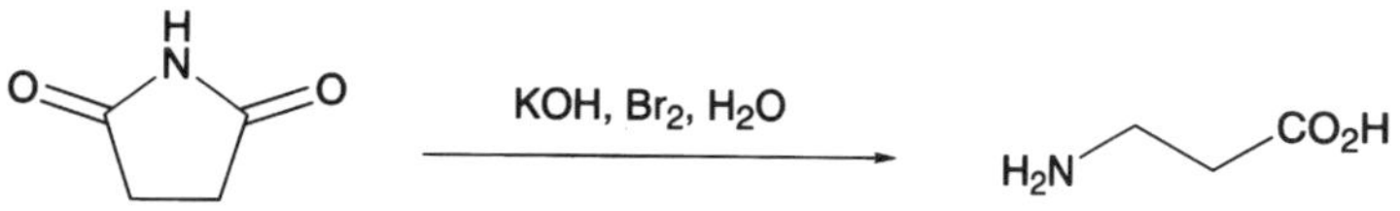

H N O O

KOH, Br_2, H_2O

H_2N CO_2H

6.17

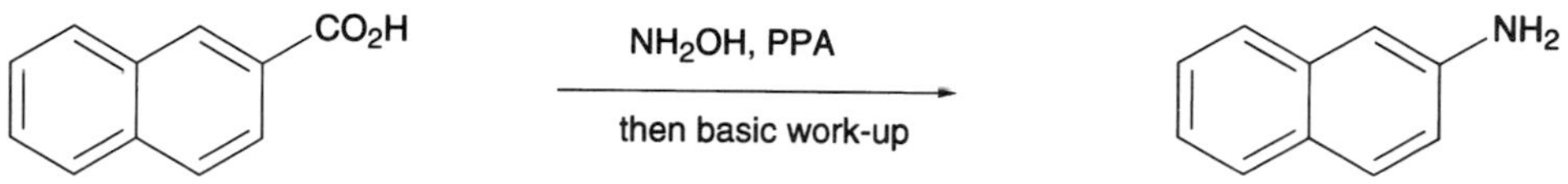

CO_2H

NH_2OH, PPA
then basic work-up

NH_2

6.18

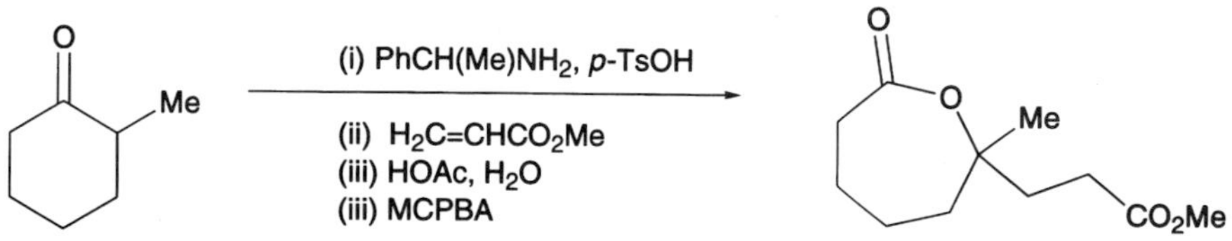

O Me

(i) PhCH(Me)NH$_2$, *p*-TsOH
(ii) H_2C=CHCO$_2$Me
(iii) HOAc, H_2O
(iii) MCPBA

O O Me CO_2Me

6.19

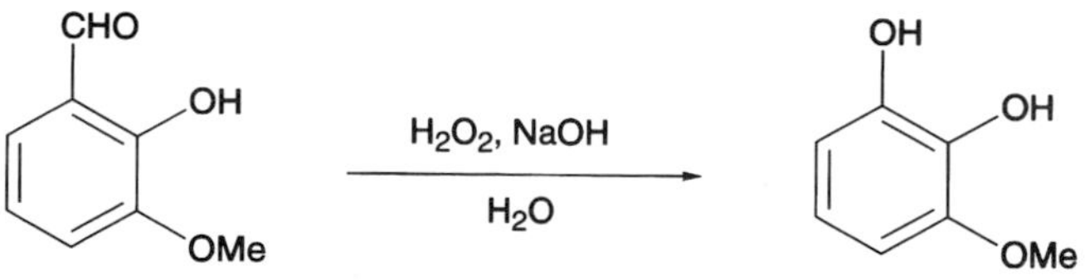

CHO OH OMe

H_2O_2, NaOH
H_2O

OH OH OMe

6.20

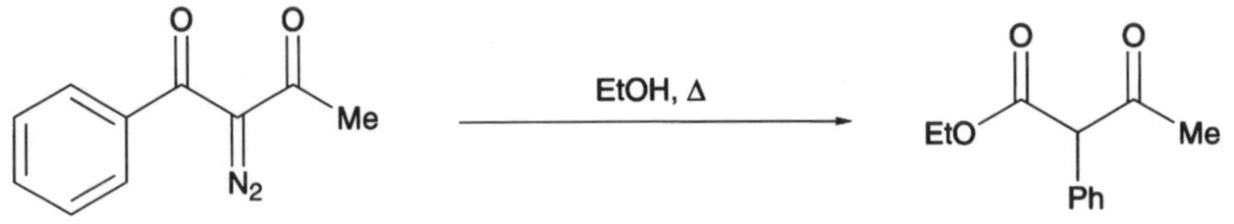

EtOH, Δ

6.21

(i) CH_2N_2
(ii) Ag_2O, EtOH
(iii) NaOH , then acidic work-up

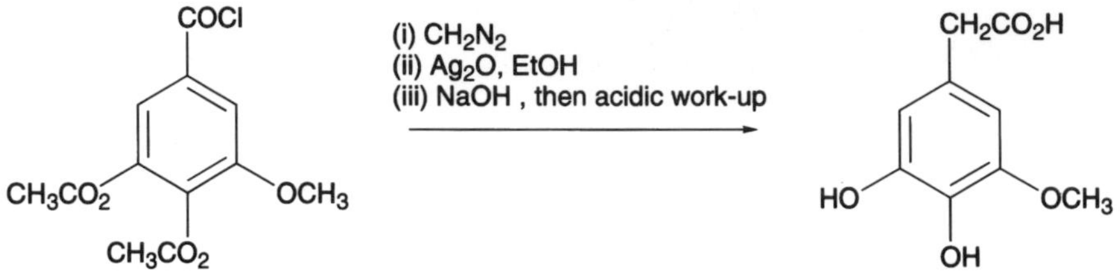

Index